In Focus – Special Book Series

Towards Circular Cities: Nature-based Solutions for Creating a Resourceful Circular City

Editors

Guenter Langergraber and Nataša Atanasova

IWA PUBLISHING

Published by **IWA Publishing**
Unit 104–105, Export Building
1 Clove Crescent
London E14 2BA, UK
Telephone: +44 (0)20 7654 5500
Fax: +44 (0)20 7654 5555
Email: publications@iwap.co.uk
Web: www.iwapublishing.com

First published 2023

British Library Cataloguing in Publication Data
A CIP catalogue record for this book is available from the British Library

ISBN: 9781789063899

Contents

doi: 10.2166/bgs.2020.001

Editorial: Towards Circular Cities – Nature-based solutions for creating a resourceful circular city

Guenter Langergraber[a] and Nataša Atanasova[b]

[a] Institute of Sanitary Engineering and Water Pollution Control, University of Natural Resources and Life Sciences, Vienna (BOKU), Muthgasse 18, A-1190 Vienna, Austria

[b] Faculty of Civil and Geodetic Engineering, University of Ljubljana, Hajdrihova 28, 1000 Ljubljana, Slovenia

'Towards Circular Cities – Nature-based solutions for creating a resourceful circular city' comprises six contributions prepared by the COST Action Circular City (https://circular-city.eu/). The Action aims to establish a network of researchers and stakeholders testing the hypothesis that: '*A circular flow system that implements nature-based solutions (NBS) for managing nutrients and resources within the urban biosphere will lead to a resilient, sustainable and healthy urban environment.*'

The six papers present the first outcome of the COST Action, triggered at the workshop, held on 13–15 February 2019 at BOKU University Vienna, Austria. They are the starting point towards deeper exploration of how NBS can contribute to circular cities, gathering state-of-the-art knowledge of one of the biggest COST Actions so far. In the first paper, we introduce the COST Action Circular City by describing its main objectives and aims as well as the general activities of the Action. The remaining five papers review the start-of-the-art of the Action's five Working Groups, i.e. Built environment, Sustainable urban water utilisation, Resource recovery, Urban Farming, and Transformation tools.

More than 100 persons from 35 COST countries contributed to the review papers. Preparing the review papers was an interdisciplinary effort in the true spirit of a COST Action network. It aimed at including the vast amount of existing knowledge in the network by first adopting a common definition of NBS and establishing a common language and understanding among disciplines. We highly appreciate the willingness and the effort of authors to contribute to this interdisciplinary work.

We do hope that 'Towards Circular Cities' will create interest for activities of the COST Action Circular City but also will help the new *Blue-Green Systems* journal to reach new readers.

doi: 10.2166/bgs.2020.933

Implementing nature-based solutions for creating a resourceful circular city

Guenter Langergraber[a,*], Bernhard Pucher[a], Lena Simperler[a], Johannes Kisser[b], Evina Katsou[c], Devi Buehler[d], Mari Carmen Garcia Mateo[e] and Nataša Atanasova[f]

[a] Institute of Sanitary Engineering and Water Pollution Control, University of Natural Resources and Life Sciences, Vienna (BOKU), Muthgasse 18, A-1190 Vienna, Austria

[b] alchemia-nova GmbH, institute for innovative phytochemistry & closed loop processes, Baumgartenstrasse 93, A-1140 Vienna, Austria

[c] Department of Civil & Environmental Engineering, Institute of Environment, Health and Societies, Brunel University London, Uxbridge Campus, Middlesex, UB8 3PH, Uxbridge, UK

[d] Institute of Natural Resource Sciences, Zurich University of Applied Sciences (ZHAW), Grüentalstrasse 22, 8820 Wädenswil, Switzerland

[e] MCG Research & Innovation Sustainability Architecture/Urban Planning, Zarandona 30004, Murcia, Spain

[f] Faculty of Civil and Geodetic Engineering, University of Ljubljana, Hajdrihova 28, 1000 Ljubljana, Slovenia

*Corresponding author. E-mail: guenter.langergraber@boku.ac.at

Abstract

Resource depletion, climate change and degradation of ecosystems are challenges faced by cities worldwide and will increase if cities do not adapt. In order to tackle those challenges, it is necessary to transform our cities into sustainable systems using a holistic approach. One element in achieving this transition is the implementation of nature-based solutions (NBS). NBS can provide a range of ecosystem services beneficial for the urban biosphere such as regulation of micro-climates, flood prevention, water treatment, food provision and more. However, most NBS are implemented serving only one single purpose. Adopting the concept of circular economy by combining different types of services and returning resources to the city, would increase the benefits gained for urban areas. The COST Action Circular City aims to establish a network testing the hypothesis that: '*A circular flow system that implements NBS for managing nutrients and resources within the urban biosphere will lead to a resilient, sustainable and healthy urban environment*.' In this paper we introduce the COST Action Circular City by describing its main objectives and aims. The paper also serves as introduction to the review papers of the Action's five Working Groups in this Special Issue.

Key words: blue-green infrastructure, circular economy, nature-based solutions, resourceful cities

INTRODUCTION

Cities worldwide are facing a number of challenges including resource depletion, climate change and degradation of ecosystems. If cities do not adapt their current infrastructure and resource management, they will not be able to cope with these challenges. Nature-based solutions (NBS) or green infrastructure (GI) solutions are one element that can help to achieve this transition. NBS and GI can provide mutual ecosystem services such as regulation of micro-climates, flood prevention, water treatment and food provision which are beneficial for the urban environment. To date, implementation of NBS focusses most of the time on achieving only one ecosystem service. The benefits gained for urban areas could be increased if the concept of circular economy is adopted by NBS achieving different ecosystem services and returning resources to the city.

The aim of the European Union-funded COST (COoperation in Science and Technology) Action Circular City is to build an interdisciplinary platform for connecting city planners, architects, system designers, circular economists, engineers and researchers from social and natural sciences that develop systems for circular management of resources in cities. Such systems would allow cities to cope with the global challenges of resource depletion as a consequence of increasing pollution and climate change. In this COST Action, the definition of a common language and understanding across disciplines are seen as crucial success factor, while circular economy (CE) concepts are seen as key approach and NBS or GI solutions are seen as core elements of the toolbox.

The COST Action aims to encourage collaboration and research to test the hypothesis that '*A circular flow system that implements NBS for managing nutrients and resources within the urban biosphere will lead to a resilient, sustainable and healthy urban environment*'. The Action tests this hypothesis in five domains: built environment, urban water, resource recovery, urban farming, and society, with particular focus on their integration to provide solutions for circular cities. It is structured according to the five domains in five Working Groups (WGs).

In this paper, we firstly define the challenge and then describe the main objectives of the COST Action Circular City as well as its structure. As starting point for the common language required, we provide key definitions that we use in the Action and, additionally, we introduce the Action's five WGs as introduction to the WG's review papers that are also part of this Special Issue.

THE CHALLENGE

Our world is approaching a situation where vital resources are peaking, e.g., oil, phosphorous, water, space, while at the same time generation of pollution is growing and climate change is proceeding. Present day's infrastructure and resource management systems are not capable of dealing with this challenge. In fact, their linearity (import–use–dispose) and consumption oriented paradigm is one of the major causes for the problems that we are facing. Continuing the 'business-as-usual' approach to resources, management will cause severe problems even in areas where such problems may seem negligible at present. Wealth and well-being of coming generations will depend on our ability to adapt our economies to this challenge in the finite world in which we are living. Transforming today's cities into sustainable and resilience cities is one of the main adaptations that will be necessary. A holistic approach looking at cities from a system's perspective is needed to achieve this goal.

NBS and especially GI are introduced in the urban landscape to cope with challenges cities are facing. These challenges are urban heat islands, flooding events, treatment of waste- and runoff waters from different origins and food provision. According to the EU Biodiversity Strategy to 2020, GI could encourage a better use of nature-based approaches to tackle climate change and to improve resource efficiency, for instance through more integrated spatial planning and development of multifunctional zones that are capable of delivering benefits to biodiversity, the land owner and society at large. NBS offer a range of ecosystem services beneficial for the environment. However, NBS are often built without considering their multifunctionality. Thus, NBS only fulfil a single function with little consideration of their recovery potential of waste and water, or their positive symbiosis with other systems. NBS can provide an array of valuable services, such as clean water production, nutrient recovery, heavy metals retention and recovery, as well as production of a broad range of

doi: 10.2166/bgs.2020.933

plant-based materials. NBS are ideally energy and resource-efficient, and resilient to change, but to be successful they must be adapted to local conditions (EC 2015a). In order to achieve successful implementation and dissemination of NBS, there is a need to raise awareness on NBS, since the concept of NBS still remains vague or unknown to the larger public (Nesshöver *et al.* 2017). Furthermore, services from NBS are often considered public goods, and their economic value is often not recognised by the markets. Consequently, their true economic worth is not reflected in society's decision-making and accounts (Kinzig *et al.* 2011).

The CE philosophy based on the 3Rs: Reduce, Reuse and Recover (EC 2014; Winans *et al.* 2017), has emerged as an alternative to the wastefulness of the current linear '*take-make-use-dispose*' practices of urban areas. The principle of CE is to create a closed loop for each natural or man-made product by transforming the linear resource flow into a circular flow. It targets all kinds of industrial processes and products. Regarding the urban environment, the scale of thinking is rather global in order to address the urban metabolism as a whole, and create not only specific CE systems, but also an overall resource management system for the urban biosphere. NBS can contribute to this on a local level as they can be easily adapted and operated decentralised where the highest demand occurs. The highest benefits of NBS besides their technical initial purposes is the influence on urban micro-climate and recreation purposes for the inhabitants.

As stated by the European Commission (EC 2015a, 2015b), CE and NBS are major parts of future developments in order to provide resources and a life-friendly environment especially but not only in urban areas. This COST Action intends to establish an interdisciplinary environment for researchers and practitioners to counter hazardous impacts of mass urbanisation and linear flows by implementing NBS. The various benefits of implementing NBS now, as described by the European Commission (2015b) can be further enhanced by the approach proposed in this COST Action. NBS contribute to sustainable urbanisation, climate change adaptation and mitigation as well as risk management and resilience. When materials become waste, the net loss of natural resources is increasing leading to the depletion of our natural capital. In addition to reducing the production and spread of hazardous materials, NBS will have additional benefits on type and method of resource use, reuse and recycling based on CE principles for the urban biosphere. This COST Action develops combined approaches to implement NBS within a CE environment enhancing the benefits provided by the implementation of NBS and increasing the reuse of 'secondary raw materials' such as organic matter, nutrients and water. This approach also represents one of the key elements for the implementation of CE (EC 2015a).

State-of-the-art

While only occupying 2% of the Earth's landscape, the urban environment consumes around 80% of the energy generated worldwide, while producing 75% of the global CO_2 emissions. The global material consumption has grown eight fold over the past 100 years and is expected to have tripled by 2050 (Krausmann *et al.* 2009; UNEP 2011; Koop & van Leeuwen 2017). With 75% of global natural resources being consumed in cities, an increasing scarcity of resources such as fertile land including nutrients, clean water and air as well as raw materials (metals, wood and plastics) is expected (EMF 2012). This generates more and more pressure on rural areas and natural ecosystems to secure the supply of water, energy, food as well as the removal of waste.

City managers and politicians are challenged to find new ways to meet these demands within their municipalities. With the concept of CE, all kinds of loops in the production chain, waste disposal and water reuse can be closed. CE adapted for cities can include the following principles: regenerate, share, optimise, loop, virtualise and exchange goods and energy (EMF 2015). In this context, different points of view concerning CE exist (EMF 2016), which are mainly dependent on the goals and mostly influenced by the involved stakeholders, and specifically addressing different materials. The overall approach stays the same, namely neglecting a linear resource flow where at the end of the chain is the disposal by closing the loop through reusing and recycling resources within a defined system. The European Commission has adopted an ambitious CE package, which encourages actions that contribute to 'closing the loop' of product lifecycles through greater recycling and reuse, with significant environmental and economic benefits (European Commission 2015b).

Despite technological innovation and improved public awareness of the environmental impacts, the increasing trend of urbanisation will make the 50% reduction in CO_2 emission by 2050 a far-fetched target. On the other hand, present day food, energy and water systems are advancing technologically, but achieving poor results when addressing the global challenges due to insufficient communication and cross-sectorial collaboration. The challenge of urban resilience is not a single sector or discipline solution. It therefore seems crucial to invest on finding interdisciplinary solutions addressing the urban metabolism as a whole pushing the frontier of the urban biosphere (Kennedy *et al.* 2009; Dong *et al.* 2016; Fujii *et al.* 2016). The benefits of NBS and GI are demonstrated in several EU-funded projects and COST Actions (e.g. COST Actions on the Green Infrastructure approach, FP1204, and on Urban Allotment Gardens, TU1201). According to the European Commission (2015a), the emerging NBS are 'living solutions inspired by, continuously supported by and using nature, which are designed to address various societal challenges in a resource-efficient and adaptable manner and to provide simultaneously economic, social, and environmental benefits'. Therefore, NBS has become a plausible concept to address the urban environmental challenges that arise as a city rapidly urbanises (EC 2015a; Maes & Jacobs 2017).

The concept of NBS builds on and supports other closely related concepts such as the ecosystem approach, which promotes the integrated management of land, water, and living resources as well as their conservation and sustainable use (Eggermont *et al.* 2015; Faivre *et al.* 2017). Moreover, the Ecosystem Services (ES) framework is used to highlight the benefits NBS provide in urban areas (Bolund & Hunhammar 1999; Haase *et al.* 2014). These benefits include local climate regulation through air cooling (Stewart & Oke 2012), mitigation of flood risks (Ozment *et al.* 2019), air pollution control (Yin *et al.* 2011; Gomez-Baggethun *et al.* 2013) and noise reduction (Bolund & Hunhammar 1999), when implementing green spaces, parks, green roofs and green walls. Direct health benefits may include positive effects on mental and physical health through stress reduction, relaxation and general health enhancements when citizens reside in urban areas (Völker & Kistemann 2011; Hartig *et al.* 2014). Finally, the presence of green and blue spaces provides the opportunity to experience nature and to enhance public ecological knowledge and awareness of nature conservation (Lundy & Wade 2011).

Finally, an important and many times overlooked service of NBS for decreasing the ecological pressure from cities is the provision of food. Edible NBS or urban agriculture systems are crucial for closing the nutrient cycle. Safely extracted resources from domestic waste flows used for urban food production address some of the biggest challenges that we are facing today: reducing waste outputs from cities, preserving decreasing phosphorous reserves by utilising phosphorus and nitrogen from wastewater, reducing food-related transportation distances and associated energy consumption and greenhouse gas emissions, and reducing the need for land- and energy-intensive food production systems. In addition, urban agriculture systems can have high levels of biodiversity, often exceeding that of other green spaces within the city, which has a positive effect on ecosystem services (Lin *et al.* 2015). Depending on the available space, different concepts of urban agriculture exist, ranging from traditional systems to vertical or underground farming as well as small decentralised systems like shared neighbourhood gardens (Bohn & Viljoen 2010; Thomaier *et al.* 2015; Buehler & Junge 2016). It has been shown that implementing different concepts can significantly increase the self-reliance of the cities (Grewal & Grewal 2011; Säumel *et al.* 2019) whereby the results of Wielemaker *et al.* (2018) reveal that integration of urban sanitation and urban agriculture can maximise urban self-sufficiency.

The shift towards more circular and sustainable modes of production and consumption is driving a shift towards greater energy efficiency and a smaller carbon footprint. In a CE, raw materials are reused and recycled; and new materials needed for the energy transition are produced more efficiently and sustainably. In turn, products are designed to be reusable, or to be easily repaired or disassembled, to facilitate remanufacturing and recycling (EPSC 2019).

It is time to systematise the use of NBS and the CE approach to resources management in cities by introducing changes in our legislation, resources utilisation concepts, technologies, economic valuation and last but not least revision of the society's values, which is a great challenge.

THE COST ACTION CIRCULAR CITY

What is a COST Action?

COST (COoperation in Science and Technology – see www.cost.eu), the longest running European framework, is a unique platform where European researchers can jointly develop their ideas and initiatives across all scientific disciplines through the trans-European networking of nationally funded research (COST 2019). COST activities are largely arranged as COST Actions. Calls for COST Actions are open, i.e. Actions can be submitted related to any scientific field.

A COST Action does not fund research. It is a science and technology network funded over a four-year duration. An Action is organised through a range of networking tools, which are performed for the purpose of supporting and ultimately achieving research coordination and capacity building objectives. Networking tools include meetings, workshops, conferences, training schools, short-term scientific missions (STSMs) and dissemination activities.

All COST activities have to be inclusive in terms of gender, age and geography. Special networking tools are available for underrepresented groups, e.g. Early Career Investigator or participants from less research-intensive countries, also referred to as Inclusiveness Target Countries.

Activities of COST Actions are coordinated by the Action Chair and Co-Chair supported by the Chairs and Co-Chairs of the Working Groups as well as the persons coordinating specific activities (e.g. STSMs and science communication). All decisions are made by the Action's Management Committee (MC). In the MC, each participating COST country can nominate two representatives. Currently, 39 countries are participating in COST and thus are COST countries. In general, activities of Actions are open to all persons working in COST countries.

For more details about COST Actions, the reader is referred to the COST website at https://www.cost.eu.

Objectives and outputs

The COST Action CA17133 Circular City provides a network for researches and practitioners from different fields: (waste-)water engineering, agronomy, urban agriculture, urban planning, architecture, energy, IT, etc., linking their fields to close knowledge gaps within the systems and scales looked at. Within this COST Action, a large number of people will be connected supporting each other and work together on finding interdisciplinary solutions to cope with the above-mentioned challenges.

The Action's objectives and outputs have been defined in its MoU. Objectives are reached with the Action's research coordination and capacity building activities, respectively. The methodology how the outputs will be delivered was not set at the beginning. Developing a common methodology is an essential part of the work in this interdisciplinary Action.

The Action's main research coordination objectives are as follows:

- Use an interdisciplinary approach applied by the different Working Groups to map occurring resources within the urban biosphere, especially provided by NBS systems.
- Develop appropriate communication methods, promoting resource recovery for consumers and built up public awareness on the benefits of a closing the loop approach.

doi: 10.2166/bgs.2020.933

- Identify, analyse and report the existing state-of-the-art of NBS implementations in the urban landscape by involving stakeholders such as city officials, urban planners and engineers.
- Identify and address regulatory, governance, financial and legal drivers and barriers for NBS implementation and use of recovered resources, and support institutional change to better regulatory governance.

The Action's main capacity building objectives are as follows:

- Widen the field of knowledge within each Working Group by incorporating a joint research approach.
- Involvement of special target groups such as city official, urban planners and engineers, gender equality and involvement of underrepresented/less research-intensive countries.
- Training of Early Career Investigators (ECI) and PhD students, in the implementation of interdisciplinary approaches concerning resource recovery, reuse and coordinating resource streams within the urban landscape when developing structural measures using NBS.

A full list of objectives can be found in the Action's Memorandum of Understanding (CA17133 2018)

The Action's main outputs are (CA17133 2018):

- a review on the state-of-the-art and existing case studies,
- a catalogue of technologies for providing/recovering resources with NBS within each WG,
- a description of possible resource input provided from NBS systems,
- scientific publications including case studies, and
- a guideline on combined NBS and CE possibilities within the urban environment.

Working Groups

The COST Action Circular Cities comprises five Working Groups. Each WG is led by a Chair and Co-Chairs elected by the MC of the Action. Besides members of the MC, also other persons can join the WGs. Usually individuals are members in only one WG for the whole duration of the Action. However, this is flexible as some individuals prefer to work in two or more WGs and/or switch WGs.

Even though each WG in the Circular City Action is focusing on a different subarea of the urban metabolism, it is necessary to connect the findings to define the potentials of embedding NBS in a CE. In order to achieve good interconnection between the individual WGs (Figure 1), the tasks and activities especially of the technology focused WGs are similar.

The Action's five WGs are as follows:

- WG1 Built environment: The built environment puts the main focus on the natural environment and its role in transforming to a CE system (Pomponi & Moncaster 2016). Within this Working Group, the NBS-CE aspect is investigated on building and settlement level with the main focus on vegetated building materials and resources to be obtained from the corresponding NBS. WG1 defines available resource streams connected to NBS within the built environment. Moreover, the aim is to identify best-practice case studies, monitor resource loops and investigate possible available resources provided by relevant NBS proposed by other WGs.
- WG2 Sustainable urban water utilisation: This Working Group investigates the implementation of a save and functional water cycle within the urban biosphere, where water is defined as a resource, nutrients can be harvested from wastewater, heavy metal adsorbed by filter materials contributing to phytomining and the treated water looped back for irrigation, sanitation and also recreational purposes. The resource recovery methods are established in WG3. WG2 critically appraises the established centralised infrastructure for water, furthermore, defines available resources within

doi: 10.2166/bgs.2020.933

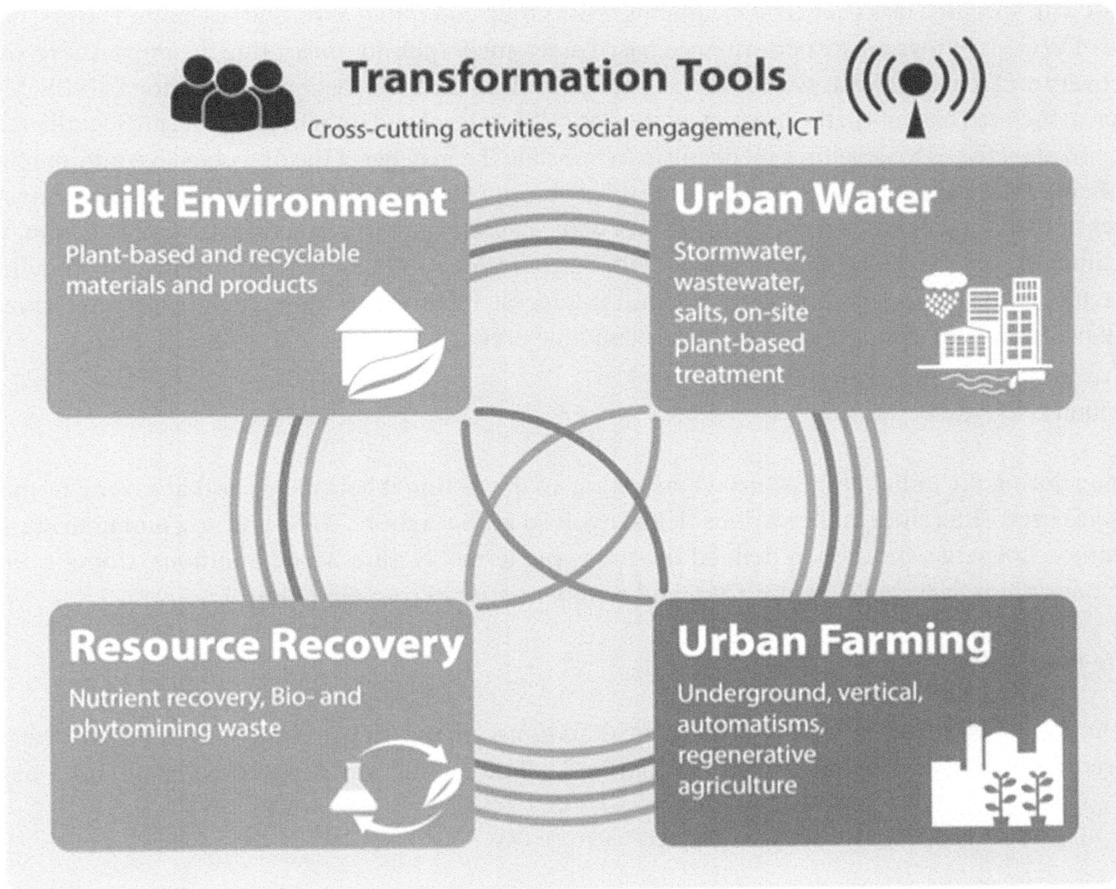

Figure 1 | The Action's Working Groups and their interrelations (CA17133 2018).

the water flow, performs risk assessment on urban water and evaluates NBS for storm water management and waste water treatment.

- WG3 Resource Recovery: A significant portion of resources is lost when passing through the urban biosphere. The implemented NBS for mitigation or treatment purposes shall become sources for a variety of resources to be harvested, used, reused and recycled within the urban environment. Therefore, possible sources and implementation strategies within the urban biosphere are investigated. WG3 identifies, appraises and assesses the available resources in the urban context, like waste- or runoff water, liquid and solid waste streams, materials from the other WGs, urban pollutants, etc. Furthermore, the aim is to combine innovative NBS applications such as regenerative biological systems, phyto- or bio-mining, bio-filtration. The outcome will be the creation of new areas for urban farming, waste treatment, runoff treatment and so on.
- WG4 Urban Farming: The WG on Urban Farming focuses on the integration of resources from other Working Groups (water, nutrients …) into urban farming systems as well as on the resources provided through urban farming for further use in other WGs. The WG will further investigate different urban farming systems especially developed for dense urban areas (e.g. underground farming, vertical and rooftop farming) and their potential for symbiosis with other WGs (e.g.: urban farming with wastewater, urban farming on formally contaminated soil after phytomining was carried out, etc.). While the main purpose of urban farming is food production within a city, the Action pays close attention to other resources available from urban farming, usually considered waste. Furthermore, the evaluated systems will consider the amount of resources available from other WGs.
- WG5 Transformation tools: WG5 coordinates and leads the interdisciplinary activities between the WGs. For this, the 'Circular City Cell' was established. The 'Cell' is composed of members from all

WGs with specified tasks aimed at facilitating cross-group communication and research. The first objective of WG5 is to investigate performance-based assessment tools for measuring the impact of resource recovery and reuse cycles as well as reviewing ICT tools to facilitate the implementation of NBS. Moreover, associated barriers to the implementation of NBS on legal and stakeholder level are identified. The second objective is to develop a mechanism to translate the insights of the first objective into simplified tools and information for stakeholders. The third objective of WG5 is to establish public relations strategies and approaches to provide stakeholders with accurate, timely and satisfactory information, with the intention to maximise public engagement. Furthermore, WG5 aims promote effective knowledge dissemination and public engagement, and to suggest methods to monitor and interpret citizens' well-being and consumption trends (socio-economic preferences).

Definitions

Already during the initial discussions when setting up the Action it became clear that several terminologies are used differently in the various fields involved in the Action. To be able to communicate with a common language, the Action defined the following terms: Nature-Based Solutions, Circular Economy, Ecosystem Services, Circular City, and micro/meso/macro-scale (Table 1).

Nature-based solutions

The Action's definition of NBS is based on definitions of the EU that refer to solutions that are inspired and supported by nature (EC 2015a) and solutions that are designed to bring more nature

Table 1 | Definition of relevant terms in the COST Action Circular City

Term	Definition
Nature-Based Solutions	NBS are defined as concepts that bring nature into cities and those that are derived from nature. NBS address societal challenges and enable resource recovery, climate mitigation and adaptation challenges, human well-being, ecosystem restoration and/or improved biodiversity status, within the urban ecosystems. As such, within this definition we achieve resource recovery using organisms (e.g. microbes, algae, plants, insects, and worms) as the principal agents. However, physical and chemical processes can be included for recovery of resources (as discussed in WG3 Resource Recovery), as they may be needed for supporting and enhancing the performance of NBS.
Ecosystem Services	Ecosystem services are the benefits people obtain from ecosystems. These ecosystem services have strong links to human well-being.
Circular Economy	CE is defined as an economic system that aims at minimising waste and making the most of resources. In a circular system resource input and waste, emission, and energy leakage are minimised by slowing, closing, and narrowing energy and material loops. This can be achieved through long-lasting design, maintenance, repair, reuse, remanufacturing, refurbishing, and recycling.
Circular City	A Circular City is where we apply the concepts of CE, i.e. we manage waste, commodities and energy in smarter and more efficient ways. A Circular City results in less pressure on our environment, new business models, innovative designs and new alliances and cooperation between different stakeholders.
Micro/meso/macro-scale	In the Action Circular City • *Micro-scale* relates to household level, • *Meso-scale* relates to district level, and • *Macro-scale* relates to city level or above. When referring to the Built Environment (WG1) • *Micro-scale* relates to green material, • *Meso-scale* relates to green buildings, whereas • *Macro-scale* relates to green sites, which are parts of the city or surrounding areas of buildings.

doi: 10.2166/bgs.2020.933

and natural features and processes to cities (EC 2015c). These EU definitions were amended by incorporating solutions that use or mimic natural processes to enhance water availability, improve water quality, and reduce risks associated with water-related disasters and climate change (UNESCO 2018) and ideas from the three criteria for NBS as suggested by Albert *et al.* (2017): First, NBS need to provide simultaneous benefits for society, the economy and nature; second, the term should be understood to represent a transdisciplinary umbrella; and third, a NBS needs to be introduced gradually.

Ecosystem services

Ecosystem services are the many and varied benefits that humans freely gain from the natural environment and from properly functioning ecosystems. Ecosystem services can be grouped into four broad categories: *provisioning*, such as the production of food and water; *regulating*, such as the control of climate and disease; *supporting*, such as nutrient cycles and oxygen production; and *cultural*, such as spiritual and recreational benefits (MEA 2005). These ecosystems services support achieving the constituents of well-being such as security, basic material for good life, health, good social relations, and freedom of choice and action (MEA 2005). NBS and treatment wetlands in cities allow to achieve multiple purposes and ecosystem services (Masi *et al.* 2018).

Circular economy

The Action's definition of CE is based on the 3Rs; Reduce, Reuse and Recover. CE is an economic system aimed at minimising waste and making the most of resources (EMF 2016, 2017). In a circular system resource input and waste, emission, and energy leakage are minimised by slowing, closing, and narrowing energy and material loops; this can be achieved through long-lasting design, maintenance, repair, reuse, remanufacturing, refurbishing, and recycling (Geissdoerfer *et al.* 2017). This regenerative approach is in contrast to the traditional linear economy, which has a 'take, make, dispose' model of production.

Circular city

A circular city is where we apply the concepts of CE, i.e. where we manage water, nutrients, commodities and energy in smarter and more efficient ways, so that minimal amount or no waste is produced. A circular city will result in less pressure on our environment, new business models, innovative designs and new alliances and cooperation between different sectors and stakeholders (Cities in Transition 2019). Alongside the UN Sustainable Development Goals (UN 2015) and climate objectives, the transition to a circular economy will support city leaders as they deliver against their priorities, which include housing, mobility, and economic development (EMF 2017).

Micro/meso/macro-scale

A general definition of micro/meso/macro-scale was not possible. Most WGs use micro for household level, meso for district level, and macro for city level or above. When referring to buildings and the built environment (WG1), the definition did not fit and was amended to micro relating to green materials, meso relating to green buildings, and macro relating to green sites, which are parts of the city or surrounding areas of buildings.

doi: 10.2166/bgs.2020.933

THIS SPECIAL ISSUE

The members of this COST Action hold a broad knowledge about different aspects of CE, NBS and Circular Cities. Therefore, the first step of the Action is to review, make the knowledge accessible to the society and create synergies. These review finally results in the Action's first main output, i.e. the 'review on the state-of-the-art and existing case studies'.

For the review of the state-of-the-art in the Action's five WGs, a common database is used as a starting point. This database comprises the available knowledge in the Action's network and includes information on 72 projects related to Circular City. Information on projects has been collected in the form of project ID cards from members of the COST Action and participants of the first Circular City workshop held in Vienna, Austria, from 14–16 February 2019.

Based on the information collected from the project ID cards, 29 out of the 72 projects are European funded and among these 29 projects, 16 are funded by European Union's Horizon 2020 Research and Innovation. Regarding the stage of the projects, 26 out of 72 are completed, while 11 projects are at an early stage of the project's development.

Among the 72 projects, 11 European and National projects are selected and presented. Table 2 provides an overview of the 11 projects, along with some general information regarding their funding source, their duration and the projects' scale (according to the definition provided in Table 1, 'In the Action Circular City'). Additionally, the projects' representatives were asked to identify the focus of their project based on the classification and allocation to the WGs of the Action.

The review of the state-of-the-art in the Action's five WGs is looking at the experiences available among Action members from different angles and identify what has been done, what was successful, what were the challenges, etc., and identify bottlenecks/research questions, as well as interlinkages between the WGs as basis for the future work in the Action. In the following, the content of the five WG state-of-the-art reviews is summarised:

- Pearlmutter *et al.* (2020) present the point of few of *WG1 Built environment*. NBS are discussed at three different levels: (i) green building materials; (ii) green building systems; and (iii) green building sites. Concepts of NBS and CE in the built environment are introduced and the impacts of urban development and the historical use of materials, systems and sites is examined. A series of case

Table 2 | General information and allocation to WGs 1–4 of 11 selected projects (EU H2020 = European Union's Horizon 2020 Research and Innovation)

Project title	Funding source	Time frame	Scale	WG 1 Built Environment	WG 2 Urban Water	WG 3 Resource Recovery	WG 4 Urban Farming
CITYFOOD	EU H2020	2018–2021	Macro	X	X	X	X
C2C-CC	European Commission	2017–2022	Macro		X		
EdiCitNet	EU H2020	2018–2023	Meso to macro	X	X	X	X
ESTIMUM	Luxembourg National Research Fund	2017–2019	Micro to macro	X	X		
HOUSEFUL	EU H2020	2018–2022	Micro	X	X	X	X
HYDROUSA	EU H2020	2018–2022	Macro			X	
Nature4Cities	EU H2020	2016–2020	Macro	X	X	X	X
Natural Water Retention Measures (NWRM)	European Commission	2013–2014	Macro		X		
proGIreg	EU H2020	2018–2023	Meso	X		X	X
Run4Life	EU H2020	2017–2021	Micro			X	
URBAN GreenUP	EU H2020	2017–2022	Micro to macro	X	X		

doi: 10.2166/bgs.2020.933

studies is presented illustrating the development and implementation of such solutions in recent years. Finally, policy instruments which can be leveraged to promote NBS and CE in the most effective manner are discussed.

- Oral *et al.* (2020) describe the *WG2 Sustainable urban water utilisation* perspective. NBS for urban water management from literature and case studies are presented and analysed. The paper identified three main challenges: (i) flood and drought protection; (ii) the water–food–energy nexus; and (iii) water purification. It is shown that NBS provide additional benefits, such as improving water quality, increasing biodiversity, obtaining social co-benefits, improving urban micro-climate, and the reduction of energy consumption by improving indoor climate. The conclusion of the paper is that NBS should be given a higher priority and should be preferred over conventional water infrastructure.
- Kisser *et al.* (2020) present the state-of-the-art review of *WG3 Resource Recovery*. The focus of the review is on NBS as technologies that bring nature into cities and those that are derived from nature, provided they enable resource recovery. The findings presented are based on an extensive literature review, as well as on original results of ongoing and recent research and innovation projects across Europe. The focus of the review was on urban wastewater, industrial wastewater, municipal solid waste and gaseous effluents, and the recoverable products (e.g. nutrients, nanoparticles, and energy). The implications of source-separation of waste and end-of-pipe technologies vs. circularity by design are discussed. Finally, an assessment of the maturity of different resource recovery technologies (Technology Readiness Level) is carried out.
- Skar *et al.* (2020) show the *WG4 Urban Farming* review. The scope of urban agriculture is to establish food production sites within the city's sphere through building-integrated agriculture including concepts such as aquaponics, indoor agriculture, vertical farming, rooftop production, edible walls, as well as through urban farms, edible landscapes, school gardens and community gardens. This article describes some of the current aspects of the circular city debate where urban agriculture is pushing forward the development of material and resource cycling in cities.
- Katsou *et al.* (2020) present the state-of-the-art in *WG5 Transformation tools*. A combined appraisal of the latest literature and a survey of projects provides an overview of enabling tools, methodologies, and initiatives for public engagement. Additionally, links between facilitators and barriers with respect to existing policies and regulations, public awareness and engagement, and scientific and technological instruments are described. The most promising methods, physical and digital technologies that may lead the way to Sustainable Circular Cities are introduced. The paper provides useful insights for citizens, scientists, practitioners, investors, policy makers, and strategists to channel efforts on switching from a linear to a circular thinking for the future of cities.

ACKNOWLEDGEMENTS

The work is funded within the COST Action CA17133 Circular City ('Implementing nature-based solutions for creating a resourceful circular city', http://www.circular-city.eu, duration 22 Oct 2018–21 Oct 2022). COST Actions are funded within the EU Horizon 2020 Programme. The authors are grateful for the support.

REFERENCES

Albert, C., Spangenberg, J. H. & Schröter, B. 2017 Nature-based solutions: criteria. *Nature* **543**, 315 (16 March 2017), https://doi.org/10.1038/543315b.

Bohn, K. & Viljoen, A. 2010 The CPUL City Toolkit: Planning productive urban landscapes for European cities. In: *Sustainable Food Planning: Evolving Theory and Practice* (Viljoen, A. & Wiskerke, J. eds). Wageningen Academic Press, Wageningen, The Netherlands.

Bolund, P. & Hunhammar, S. 1999 Ecosystem services in urban areas. *Ecological Economics* **29** (2), 293–301.

Buehler, D. & Junge, R. 2016 Global trends and current status of commercial urban rooftop farming. *Sustainability* **8** (11), 1108. doi:10.3390/su8111108.

CA17133 2018 Memorandum of Understanding for the Implementation of the COST Action 'Implementing Nature Based Solutions for Creating A Resourceful Circular City'. Brussels, Belgium. Available from: https://e-services.cost.eu/files/domain_files/CA/Action_CA17133/mou/CA17133-e.pdf (accessed 30 May 2019).

Cities in Transition 2019 *The Circular City*. Available from: https://citiesintransition.eu/transition/circular-city (accessed 30 May 2019).

COST 2019 *COST Vademecum*. COST office, Brussels, Belgium. Available from: https://www.cost.eu/wp-content/uploads/2019/05/Vademecum-May-2019.pdf (accessed 30 May 2019).

Dong, H., Fujita, T., Geng, Y., Dong, L., Ohnishi, S., Sun, L., Dou, Y. & Fujii, M. 2016 A review on eco-city evaluation methods and highlights for integration. *Ecological Indicators* **60**, 1184–1191.

EC 2014 *Scoping Study to Identify Potential Circular Economy Actions, Priority Sectors, Material Flows & Value Chains*. European Commission, Brussels, Belgium.

EC 2015a Towards an EU Research and Innovation Policy Agenda for Nature-Based Solutions & Re-Naturing Cities. *Final Report of the Horizon 2020 Expert Group on Nature-Based Solutions and Re-Naturing Cities (full version)*; Publications Office of the European Union, European Commission, und Directorate-General for Research and Innovation, Luxembourg. Available from: http://dx.publications.europa.eu/10.2777/765301 (accessed 30 May 2019).

EC 2015b *Closing the Loop – An EU Action Plan for the Circular Economy*. European Commission, Brussels, Belgium.

EC 2015c Available from: https://ec.europa.eu/research/environment/index.cfm?pg=nbs (accessed 30 May 2019).

Eggermont, H., Balian, E., Azevedo, J. M. N., Beumer, V., Brodin, T., Claudet, J. & Reuter, K. 2015 Nature-based solutions: new influence for environmental management and research in Europe. *GAIA – Ecological Perspectives for Science and Society* **24** (4), 243–248.

EMF 2012 *Towards the Circular Economy: Economic and Business Rationale for Accelerated Transition*. Ellen MacArthur Foundation, Cowes, UK. Available from: https://www.ellenmacarthurfoundation.org/assets/downloads/TCE_Ellen-MacArthur-Foundation_9-Dec-2015.pdf (accessed 30 May 2019).

EMF 2015 *Delivering the Circular Economy: A Toolkit for Policymakers*. Ellen MacArthur Foundation, Cowes, UK. Available from: https://www.ellenmacarthurfoundation.org/assets/downloads/publications/EllenMacArthurFoundation_PolicymakerToolkit.pdf (accessed 30 May 2019).

EMF 2016 *Circular Economy – Schools of Thought*. Ellen MacArthur Foundation, Cowes, UK. Available from: https://www.ellenmacarthurfoundation.org/circular-economy/concept/schools-of-thought (accessed 30 May 2019).

EMF 2017 *Circular Economy in Cities*. Ellen MacArthur Foundation, Cowes, UK. Available from: https://www.ellenmacarthurfoundation.org/our-work/activities/circular-economy-in-cities (accessed 30 May 2019).

EPSC 2019 *European Political Strategy Centre*. Available from: https://ec.europa.eu/epsc/home_en (accessed 30 May 2019).

Faivre, N., Fritz, M., Freitas, T., de Boissezon, B. & Vandewoestijne, S. 2017 Nature-Based Solutions in the EU: innovating with nature to address social, economic and environmental challenges. *Environmental Research* **159**, 509–518.

Fujii, M., Fujita, T., Dong, L., Lu, C., Geng, Y., Behera, S. K., Park, H.-S. & Chiu, A. S. F. 2016 Possibility of developing low-carbon industries through urban symbiosis in Asian cities. *Journal of Cleaner Production* **114**, 376–386.

Geissdoerfer, M., Savaget, P., Bocken, N. M. P. & Hultink, E. J. 2017 The circular economy – a new sustainability paradigm? *Journal of Cleaner Production* **143**, 757–768.

Gomez-Baggethun, E., Gren, Å, Barton, D. N., Langemeyer, J., McPherson, T., O'Farrell, P., Andersson, E., Hamsted, Z. & Kremer, P. 2013 Urban ecosystem services. In: *Urbanization, Biodiversity and Ecosystem Services: Challenges and Opportunities. A Global Assessment* (Elmqvist, T., Fragkias, M., Goodness, J., Güneralp, B., Marcotullio, P. J., McDonald, R. I., Parnell, S. & Schewenius, M. eds). Springer, Dordrecht, The Netherlands, pp. 175–251.

Grewal, S. S. & Grewal, P. S. 2011 Can cities become self-reliant in food? *Journal of Cities* **29**, 1–11.

Haase, D., Larondelle, N., Andersson, E., Artmann, M., Borgström, S., Breuste, J., Gomez-Baggethun, E., Gren, Å., Hamstead, Z., Hansen, R., Kabisch, N., Kremer, P., Langemeyer, J., Lorance, E., McPhearson, T., Pauleit, S., Qureshi, S., Schwarz, N., Voigt, A., Wurster, D. & Elmqvist, T. 2014 A quantitative review of urban ecosystem service assessments: concepts, models, and implementation. *AMBIO* **43** (4), 413–433.

Hartig, T., Mitchell, R., de Vries, S. & Frumkin, H. 2014 Nature and health. *Annual Review of Public Health* **35**, 207–228.

Katsou, E., Nika, C.-E., Buehler, D., Marić, B., Megyesi, B., Mino, E., Babí Almenar, J., Bas, B., Bećirović, D., Bokal, S., Đolić, M., Elginöz, N., Kalnis, G., Garcia Mateo, M. C., Milousi, M., Mousavi, A., Rinčić, I., Rizzo, A., Rodriguez-Roda, I., Rugani, B., Šalaševičienė, A., Sari, R., Stanchev, P., Topuz, E. & Atanasova, N. 2020 Transformation tools enabling the implementation of nature-based solutions for creating a resourceful circular city. *Blue-Green Systems*. https://doi.org/10.2166/bgs.2020.929.

Kennedy, C., Steinberger, J., Gasson, B., Hansen, Y., Hillman, T., Havránek, M., Pataki, D., Phdungsilp, A., Ramaswami, A. & Mendez, G. V. 2009 Greenhouse gas emissions from global cities. *Environmental Science and Technology* **43**, 7297–7302.

Kinzig, A. P., Perrings, C., Chapin, F. S., Polasky, S., Smith, V. K., Tilman, D. & Turner, B. L. 2011 Paying for ecosystem services – promise and peril. *Science* **334** (6056), 603–604.

Kisser, J., Wirth, M., De Gusseme, B., Van Eekert, M., Zeeman, G., Schoenborn, A., Vinnerås, B., Finger, D. C., Kolbl Repinc, S., Griessler Bulc, T., Bani, A., Pavlova, D., Staicu, L. C., Atasoy, M., Cetecioglu, Z., Kokko, M., Haznedaroglu, B. Z., Hansen,

J., Istenič, D., Canga, E., Malamis, S., Camilleri-Fenech, M. & Beesley, L. 2020 A review of nature-based solutions for resource recovery in cities. *Blue-Green Systems* 2 (1), 138–172. https://doi.org/10.2166/bgs.2020.930.

Koop, S. H. A. & van Leeuwen, C. J. 2017 The challenges of water, waste and climate change in cities. *Environment, Development and Sustainability* **19** (2), 385–418.

Krausmann, F., Gingrich, S., Eisenmenger, N., Erb, K.-H., Haberl, H. & Fischer-Kowalski, M. 2009 Growth in global materials use, GDP and population during the 20th century. *Ecological Economics* **68** (10), 2696–2705.

Lin, B. B., Philpott, S. M. & Jha, S. 2015 The future of urban agriculture and biodiversity-ecosystem services: challenges and next steps. *Basic and Applied Ecology* **16** (3), 189–201.

Lundy, L. & Wade, R. 2011 Integrating sciences to sustain urban ecosystem services. *Progress in Physical Geography* **35**, 653–669.

Maes, J. & Jacobs, S. 2017 Nature-based solutions for Europe's sustainable development. *Conservation Letters* **10** (1), 121–124.

Masi, F., Rizzo, A. & Regelsberger, M. 2018 The role of constructed wetlands in a new circular economy, resource oriented, and ecosystem services paradigm. *Journal of Environmental Management* **216**, 275–284.

MEA 2015 *Ecosystems and Human Well-Being*. Synthesis, Millenium Esosystems Assessment, Island Press, Washington, DC, USA. Available from: https://www.millenniumassessment.org/documents/document.356.aspx.pdf (accessed 30 May 2019).

Nesshöver, C., Assmuth, T., Irvine, K. N., Rusch, G. M., Waylen, K. A., Delbaere, B., Haase, D., Jones-Walters, L., Keune, H., Kovacs, E., Krauz, K., Külvik, M., Rey, F., van Dijk, J., Vistad, O. I., Wilkinson, M. E. & Wittmer, H. 2017 The science, policy and practice of nature-based solutions: an interdisciplinary perspective. *Science of The Total Environment* **579**, 1215–1227.

Oral, H. V., Carvalho, P., Gajewska, M., Ursino, N., Masi, F., van Hullebusch, E. D., Kazak, J. K., Exposito, A., Cipolletta, G., Andersen, T. R., Finger, D. C., Simperler, L., Regelsberger, M., Rous, V., Radinja, M., Buttiglieri, G., Krzeminski, P., Rizzo, A., Dehghanian, K., Nikolova, M. & Zimmermann, M. 2020 A review of nature-based solutions for urban water management in European circular cities: a critical assessment based on case studies and literature. *Blue-Green Systems* 2 (1), 112–136. https://doi.org/10.2166/bgs.2020.932.

Ozment, S., Ellison, G. & Jongman, B. 2019 *Nature-Based Solutions for Disaster Risk Management: Booklet (English)*. World Bank Group, Washington, DC, USA. Available from: http://documents.worldbank.org/curated/en/253401551126252092/Booklet (accessed 22 October 2019).

Pearlmutter, D., Theochari, D., Nehls, T., Pinho, P., Piro, P., Korolova, A., Papaefthimiou, S., Garcia Mateo, M. C., Calheiros, C., Zluwa, I., Pitha, U., Schosseler, P., Florentin, Y., Ouannou, S., Gal, E., Aicher, A., Arnold, K., Igondová, E. & Pucher, B. 2020 Enhancing the circular economy with nature-based solutions in the built urban environment: Green building materials, systems and sites. *Blue-Green Systems* 2 (1), 46–72. https://doi.org/10.2166/bgs.2019.928.

Pomponi, F. & Moncaster, A. 2016 Circular economy for the built environment: a research framework. *Journal of Cleaner Production* **143**, 710–718.

Säumel, I., Reddy, S. E. & Wachtel, T. 2019 Edible city solutions – one step further to foster social resilience through enhanced socio-cultural ecosystem services in cities. *Sustainability* **11**, 972.

Skar, S. L. G., Pineda-Martos, R., Timpe, A., Pölling, B., Bohn, K., Külvik, M., Delgado, C., Pedras, C. M. G., Paço, T. A., Ćujić, M., Tzortzakis, N., Chrysargyris, A., Peticila, A., Alencikiene, G., Monsees, H. & Junge, R. 2020 Urban agriculture as an innovative mechanism to secure cities a sustainable and healthy development in the future. *Blue-Green Systems* 2 (1), 1–27. https://doi.org/10.2166/bgs.2019.931.

Stewart, I. D. & Oke, T. R. 2012 Local climate zones for urban temperature studies. *Bulletin of the American Meteorological Society* **93**, 1879–1900.

Thomaier, S., Specht, K., Henckel, D., Dierich, A., Siebert, R., Freisinger, U. B. & Sawicka, M. 2015 Farming in and on urban buildings: present practice and specific novelties of zero-acreage farming (ZFarming). *Renewable Agriculture and Food Systems* **30**, 43–54.

UN 2015 *UN Sustainable Development Goals*. Available from: https://sustainabledevelopment.un.org/?menu=1300 (accessed 30 May 2019).

UNEP 2011 *Towards A Green Economy: Pathways to Sustainable Development and Poverty Eradication – A Synthesis for Policy Makers*. UNEP. Available from: https://sustainabledevelopment.un.org/content/documents/126GER_synthesis_en.pdf (accessed 30 May 2019).

UNESCO 2018 Nature-based Solutions for Water. In: *World Water Development Report 2018*. UNESCO, Paris, France. Available from: https://unesdoc.unesco.org/ark:/48223/pf0000261424 (accessed 30 May 2019).

Völker, S. & Kistemann, T. 2011 The impact of blue space on human health and well-being – salutogenetic health effects of inland surface waters: a review. *International Journal of Hygiene and Environmental Health* **214** (6), 449–460.

Wielemaker, R. C., Weijma, J. & Zeeman, G. 2018 Harvest to harvest: recovering nutrients with new sanitation systems for reuse in urban agriculture. *Resources, Conservation and Recycling* **128**, 426–437.

Winans, K., Kendall, A. & Deng, H. 2017 The history and current applications of the circular economy concept. *Renewable and Sustainable Energy Reviews* **19** (2), 385–418.

Yin, S., Shen, Z., Zhou, P., Zou, X., Che, S. & Wang, W. 2011 Quantifying air pollution attenuation within urban parks: an experimental approach in Shanghai, China. *Environmental Pollution* **159**, 2155–2163.

First received 27 June 2019; accepted in revised form 15 January 2020. Available online 4 March 2020

doi: 10.2166/bgs.2019.928

Enhancing the circular economy with nature-based solutions in the built urban environment: green building materials, systems and sites

David Pearlmutter[a,*], Dimitra Theochari[b], Thomas Nehls[c], Pedro Pinho[d], Patrizia Piro[e], Alisa Korolova[f], Spiros Papaefthimiou[g], Mari Carmen Garcia Mateo[h], Cristina Calheiros[i], Irene Zluwa[j], Ulrike Pitha[j], Paul Schosseler[k], Yaakov Florentin[l], Shahar Ouannou[l], Erez Gal[m], Andreas Aicher[n], Killian Arnold[o], Erika Igondová[p] and Bernhard Pucher[q]

[a] Department of Geography and Blaustein Institutes for Desert Research, Ben-Gurion University of the Negev, Beer Sheva, Israel

[b] Ramboll Studio Dreiseitl/ MSB Landschaftsarchitekten; PhD Cand, National Technical University of Athens, Stournari and Patision, 42 35, Athina 106 82, Greece

[c] Center for Innovation and Science on Building Greening (CIBG) Chair for Ecohydrology, Institute of Ecology Technische Universitaet, Berlin, Germany

[d] cE3c-Centre for Ecology, Evolution and Environmental Changes, Faculdade de Ciências da Universidade de Lisboa, Portugal

[e] Department of Civil Engineering, University of Calabria, Italy

[f] Faculty of Architecture, Riga Technical University, Latvia

[g] School of Production Engineering & Management, Technical University of Crete, Greece

[h] MCG Research & Innovation Sustainability Architecture/Urban Planning, Spain

[i] Interdisciplinary Centre of Marine and Environmental Research (CIIMAR/CIMAR), University of Porto, Portugal

[j] Institute of Soil Bioengineering and Landscape Construction, University of Natural Resources and Life Sciences Vienna (BOKU), Austria

[k] PositiveImpaKT Solarwind Windhof, Luxembourg

[l] Blaustein Institutes for Desert Research, Ben-Gurion University of the Negev, Israel

[m] Department of Structural Engineering, Ben-Gurion University of the Negev, Israel

[n] Urban Water Management and Sanitation, Bauhaus Institute for Infrastructure Solutions, Bauhaus Universitaet Weimar, Germany

[o] Engineering and Architecture, Lucerne University of Applied Sciences and Arts, Switzerland

[p] Department of Urban Planning, The Municipality of the Capital City of Bratislava, Slovakia

[q] Institute of Sanitary Engineering and Water Pollution Control, University of Natural Resources and Life Sciences Vienna (BOKU), Austria

*Corresponding author. E-mail: davidp@bgu.ac.il

Abstract

The objective of this review paper is to survey the state of the art on nature-based solutions (NBS) in the built environment, which can contribute to a circular economy (CE) and counter the negative impacts of urbanization

doi: 10.2166/bgs.2019.928

through the provision of ecosystem services. NBS are discussed here at three different levels: (i) green building materials, including biocomposites with plant-based aggregates; (ii) green building systems, employed for the greening of buildings by incorporating vegetation in their envelope; and (iii) green building sites, emphasizing the value of vegetated open spaces and water-sensitive urban design. After introducing the central concepts of NBS and CE as they are manifested in the built environment, we examine the impacts of urban development and the historical use of materials, systems and sites which can offer solutions to these problems. In the central section of the paper we present a series of case studies illustrating the development and implementation of such solutions in recent years. Finally, in a brief critical analysis we look at the ecosystem services and disservices provided by NBS in the built environment, and examine the policy instruments which can be leveraged to promote them in the most effective manner – facilitating the future transition to fully circular cities.

Key words: built environment, circular economy, nature-based solutions

INTRODUCTION

As defined by Langergraber *et al.* (2019), nature-based solutions (NBS) are concepts that bring nature into cities – and in many cases this includes ideas for urban design that are inspired or derived from nature. Thus while NBS may be considered more generally as actions which protect, sustainably manage, and restore natural or modified ecosystems (IUCN 2019), the specific focus here is on the implementation of NBS within *urban* ecosystems.

Even within cities, NBS contribute to global objectives such as climate change mitigation and adaptation, and they have the potential to enhance human wellbeing biodiversity and resource recovery. All of these goals find expression in the design of buildings and urban spaces, or what is commonly referred to as the 'built environment'. Over the last 15 years, in fact, the concept of NBS has come to encompass design solutions for contemporary landscapes and architecture, in which natural and living material – as well as policies, measures and actions promoting their use – are leveraged to meet specific societal challenges that are pervasive in the built environment.

One example of a societal challenge that may addressed using NBS is the urban heat island effect (UHI), by which temperatures in cities are increasingly higher than in surrounding areas – exacerbating heat stress for vulnerable urban populations. In temperate climates this risk is highest during the night time, and mainly indoors (Buchin *et al.* 2016). Among UHI countermeasures, urban green space is considered among the most effective for reducing air temperatures outdoors, but for addressing the indoor hazard it is most effective to apply measures at the building level – highlighting the fact that NBS must be implemented at a range of scales in order to deliver the optimum benefit (Saaroni *et al.* 2018).

The ubiquitous grey infrastructure in cities – consisting of impervious paving, buildings, and other structures – contributes as a whole to the worsening of the urban climate through a lack of resilience and flexibility. Thus the importance of urban green infrastructure (GI) is related to the amelioration of social stresses, which are intertwined with physical phenomena in cities. Contemporary issues related to grey infrastructure include its ageing and the need for maintenance, along with the worldwide recognition that conventional infrastructure solutions are often insufficient and ineffective. The inclusion of NBS in new and innovative strategies can address such issues as water quality while simultaneously delivering additional benefits. These benefits are vital to promote aspects of sustainable development presented in a 2018 United Nations report which states that '…upscaling (of) NBS will be central to achieving the 2030 Agenda for Sustainable Development' (WWAP 2018).

Another important aspect when approaching contemporary urban systems is 'circularity'.

Circular economy (CE) is an evolving 'umbrella' concept embodying internal complexities and multiple definitions, but is defined here (Langergraber *et al.* 2019) as an economic system that aims at minimising waste and making the most of resources. In a circular system, resource input and waste, emission, and energy leakage are minimized by slowing, closing, and narrowing energy and material loops. According to Prieto-Sandoval *et al.* (2018), 'this concept represents a change of paradigm in the way that human society is interrelated with nature and aims to prevent the depletion of resources and facilitate sustainable development through its implementation at the micro (enterprises and consumers), meso (economic agents integrated in symbiosis) and macro (city, regions and governments) levels. Attaining this circular model requires cyclical and regenerative environmental innovations in the way society legislates, produces and consumes.'

Moreover, cities are composed of multiple overlapping infrastructure systems, while at the same time become the spaces where people develop social experiences and cultural values. Urban quality is a target for cities that are competing against each other to attract the most valuable and entrepreneurial citizens. A great number of theories, manifestos and city guidelines have been developed in cities of the world to promote new urban qualities and citizen wellbeing. Frederick Law Olmsted argued that '...the enjoyment of scenery employs the mind without fatigue and yet exercises it, tranquilizes it and yet enlivens it; and thus, through the influence of the mind over the body, gives the effect of refreshing rest and reinvigoration to the whole system' (Olmsted 1865).

One of the central concepts underlying NBS which has been recognized for several decades by the scientific and design communities, and intuitively for hundreds of years by the population at large, is *biophilia* – defined as the 'innate human attraction to nature'. Biophilic design is of growing interest within architectural theory and practice, due to published research showing that design which increases the exposure and direct connection of people with the natural world 'can reduce stress, improve cognitive function and creativity, improve our well-being and expedite healing' (Browning *et al.* 2014). Social life and culture develop constantly in the city context and their relation to nature and new green infrastructure solutions is a very important goal for NBS.

Against this background, the aim of this review paper is to survey the state of the art on integrating NBS in the built environment, which can counter the negative impacts of urbanization and contribute to a circular economy through the provision of ecosystem services. We approach this review at three scales of implementation, those of building *materials*, *systems*, and *sites* (Figure 1) – which, as defined below, constitute the focus of the paper.

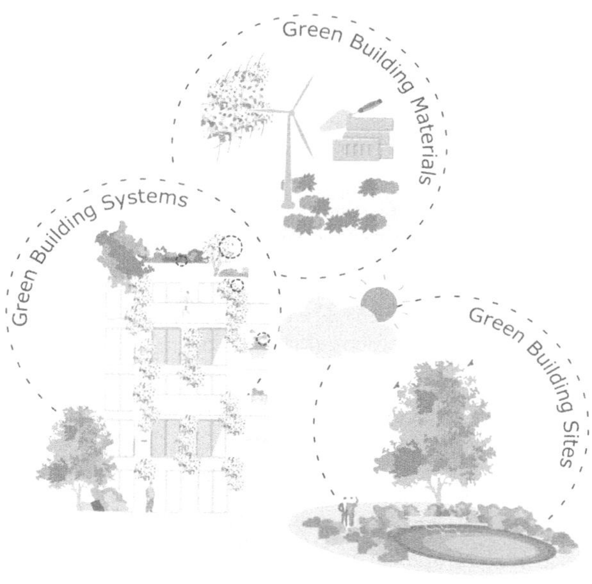

Figure 1 | Three scales of NBS implementation in the built environment: green building materials, systems for the greening of buildings, and green urban sites. Illustration: Dimitra Theochari (unauthorized use is not permitted).

doi: 10.2166/bgs.2019.928

Green building materials are raw and processed nature-based materials used in the construction of the built environment. These materials are extracted from the biological cycle to serve technical purposes, and their production and processing should result in low environmental impacts in terms of measures such as embodied energy and carbon, water consumption and the use of harmful chemicals. Ideally, they make productive reuse of other resource streams to avoid detrimental by-products and competition with food production, and they guarantee a healthy working and living environment with respect to indoor air quality and climate. The material processing and construction techniques should ideally be such that nutrients can be safely returned to the ecosystem at the end of the use cycle of the building material.

Green building systems in this context are systems for the greening of buildings, and include components such as green roofs, façade greenery and living walls, house trees, and even building-integrated constructed wetlands. Green roofs are designed with either intensive or extensive planting: an intensive green roof is supplied with water and nutrients and its substrate is usually thicker than 0.25 m, while an extensive green roof is not irrigated and has a much shallower substrate, usually 0.06–0.15 m. Façade greenery consists of clingers or climbers rooting in soil or artificial substrates, directly attaching to the wall surface or covering the wall indirectly through support systems like trellises or ropes. Since the plants use other structures to develop, they do not have to invest in their own static apparatus – an evolutionary adaptation that often allows them to grow much faster than trees (e.g. *Fallopia baldshuanica* reaches up to 12 m in height in just one growing season).

Green building sites may be open spaces directly adjacent to buildings, typically within in the same property, or land parcels of small and medium scale (pocket parks, urban plazas, small community parks, elevated urban green promenades) that have a role in the blue-green (i.e. water and vegetation-based) network of the city. Green building sites are spaces for establishing nature in cities, enhancing biodiversity through blue-green infrastructure components, providing opportunities for biophilic design and promoting culture and social life through activities for diverse social and age groups. Ideally green building sites provide multiple ecosystem services, and embody resilient and regenerative ways to deal simultaneously with challenges ranging from climate change mitigation to the reduction of noise pollution. They may increase flood safety through integrated water sensitive urban design (WSUD) solutions, improve air quality, reuse site material while reducing construction waste, use construction materials that are permeable and sustainable, promote local economic systems, and maintain low life cycle impacts. The intent of NBS at the green building sites scale is to promote sustainability goals, outdoor comfort, healthy living environments and wellbeing in cities.

NBS IN THE BUILT ENVIRONMENT: PROBLEMS AND SOLUTIONS

Meeting the challenges posed by urbanization and the growth of cities has a pivotal role to play in the transition of society to a circular economy (CE). These challenges derive from a wide variety of environmental impacts associated with the production of building materials, the operation of buildings and their allied systems, and the outdoor processes which take place within the built-up area (Elmquist *et al.* 2013). In this section we first summarize these impacts, and then survey the historical use of nature-based solutions that can be leveraged in the future to minimize their negative consequences.

Impacts of the built environment on people and natural systems

In Europe, energy use in residential and commercial buildings accounts for over 40% of the total end-use consumption (Enerdata 2012), making buildings the largest energy-consuming sector and a major emitter of greenhouse gases (GHG). However, this proportion refers only to 'operational' energy use, or that which is consumed for heating, cooling, ventilation, and otherwise making the building habitable and supporting the ongoing needs of its occupants. Additional energy consumption and related GHG emissions are attributable to the production of the building and its constituent materials, and to the characteristics of urban open spaces – in which microclimatic and other processes crucially influence the dependence of people on energy-intensive buildings and urban transport. Therefore, the design of the built environment has a determinative impact not only on the quality of contemporary urban life, but also on society's long-term contribution to anthropogenic climate change and natural resource depletion (Kabisch *et al.* 2017).

Building material production

Most modern building materials are produced using processes that rely heavily on non-renewable resources, generate large quantities of waste, and potentially create unhealthy surroundings. The most ubiquitous of these materials is concrete, which may be either cast on site or used to manufacture pre-cast elements like wall blocks, and whose key active ingredient is Portland cement. Cement has an extremely high level of embodied energy in its production, and is a major source of GHG emissions (Huberman *et al.* 2015). This is because vast quantities of carbon dioxide are released to the atmosphere as part of the calcination process in which natural limestone and other raw materials are converted into clinker (the key ingredient in cement), and additional CO_2 is emitted due to the combustion of fossil fuel for producing the required temperature of approximately 1,500 °C. Because of such high-temperature production processes, many other modern materials have high embodied energy as well. These include the reinforcing used in structural concrete (especially when it is virgin steel produced from iron ore, rather than recycled from scrap), and even materials that are increasingly employed for thermal energy efficiency – such as aluminium and glass for high-performance windows, and petrochemical-based plastics such as expanded polystyrene for thermal insulation. Beyond embodied energy and carbon emissions, a host of direct and indirect offsite environmental impacts (such as air pollution and water contamination) are also incurred through the extraction of minerals from mines and quarries, and the long-distance transport of both raw materials and finished products.

The building and its operational systems

Buildings provide people with shelter from cold, heat, wind and precipitation, but also have negative impacts on organisms and ecosystems. Among the direct impacts are the reduction of vegetation and disconnection of habitats on the ground, as buildings seal the soil and thus disrupt the cycles of water, gases, nutrients and energy. On building roofs, rainwater is concentrated in volume and time, and about 85% is typically directed to drainage where it infiltrates directly into the ground or into sewers – bypassing soils, plants and the atmosphere. Buildings impact the urban surface roughness and turbulent exchanges of heat and pollutants, often trapping them in streets and other occupied spaces. Compared to a vegetated surface, buildings also increase the absorptive surface and the heat storage capacity by accumulation of high density, high heat capacity materials, which leads to longwave radiation emission during the night and the formation of urban heat islands – contributing to heat stress both indoors and outdoors. In cold northern climates, densely packed buildings can block sunlight to an undesirable extent in terms of both health and potential passive solar heating.

Because urban populations spend the large majority of their time indoors, their consumption of water, materials and energy – and their production of wastewater, waste and excess (anthropogenic) heat – are concentrated within buildings. Biological primary production is reduced to a minimum, as food is imported, processed and consumed, with residues and waste flushed into the sewer

doi: 10.2166/bgs.2019.928

system – again bypassing the soil (where these organic materials are usually decomposed and mineralized) and the vegetation (which is usually taking up the corresponding nutrients). Heating and cooling in homes, businesses and industry consume around half of the energy produced in the EU (Enerdata 2012), and buildings are projected to remain the largest energy-use sector worldwide, even under future decarbonisation scenarios (US EIA 2017).

Outdoor processes in building sites

Urban spaces, typically covered by hard paving using materials like concrete and asphalt, can create environmental problems both within and beyond their boundaries. Without shade from trees, such spaces exacerbate thermally stressful microclimatic conditions on hot days and discourage pedestrian activity – which can in turn increase the reliance of city dwellers on air-conditioned vehicles and indoor spaces. When the albedo of unplanted ground surfaces is low, their surface temperature may reach extreme levels, and when it is high they expose users to reflected solar radiation which increases visual as well as thermal stress. Hydrologically, large areas of impervious paving contribute to surface runoff that may lead to soil erosion, impaired water quality and the risk of flooding.

These sites also diminish biodiversity, by creating a harsh environment for wildlife. The strongest impact comes from direct habitat destruction, when existing green spaces are replaced by buildings and roads. Remaining habitats, as well as newly built ones, also face challenges, mostly related to habitat fragmentation. Green spaces in cities can be viewed as islands isolated by a hostile environment, and thus species present must cope with different abiotic pressures, e.g. from increased air pollution (mostly from vehicles), altered microclimate (related to the urban heat-island effect) and isolation (which limits species dispersion between green spaces). Also, the biotic interactions between species are affected due to the presence of exotic and invasive species, and the prevalence of disturbances stemming from human activities, such as noise pollution.

Historical overview of NBS in the built environment

Traditional building materials and techniques

The use of 'nature-based solutions' in the built environment is not a new phenomenon, as builders throughout history have employed materials and techniques enabled by the natural surroundings. Indigenous architecture around the world is characterized by buildings that were constructed from local materials and often displayed a remarkable unity with their environment (Blaser 1982). By using locally available resources in ways that respond to local conditions, traditional building practice has led to the evolution of distinct regional building types. Prominent examples vary from domed igloos built of ice and snow in arctic regions to solid earth construction (cob, rammed earth, mud bricks) without an interior skeleton structure, as can be seen in adobe Pueblos of the arid American southwest (Lehner 2016). Natural stone is among the most important building materials due to its strength and durability, either forming whole huts (Italian Trulli or Bories in the Provence) or building up houses with thick stone walls supporting a turf roof with living grass sod (Blaser 1982). As detailed below, living plant material was also used to form ficus tree bridges in India (Rodgers 2019).

More common than living plant material is the use of harvested material, in a wide range of construction types from the compact wooden architecture of Europe to light bamboo piles in Asia. Tipis of the Great Plains and yurts of Central Asia exemplify skeleton structures of wood or other plant material covered by tensile fabrics of plant-fiber or animal skins. The Hawaiian hale consists of a more complex skeleton structure system, containing a ridge purlin and different junctions, covered with plant leaves. In the Nile Valley of Egypt, typical dwellings were constructed using an inner structure made of wooden poles, reeds or wickerwork, covered by a layer of mud. Another plant-based material typical for eastern construction is paper, used for Japanese shoji doors or fusuma screens (Brown 2012).

The use of wood in construction traditionally reflects the proximity of forests and the availability of tree species with given properties. In areas such as northern Europe that are rich in trees with hard and solid stems, heavy timber walls and snow-bearing pitched roofs with wooden planks have been historically pervasive – though today they have largely disappeared from many regions due to the high price of wood. In areas where forest trees yield thinner and shorter trunks, there is a more common use of lighter wood-frame walls filled with other materials such as brick, and where timber is scarce, soil is often used as a construction material, together with light wood and reed. In stony-karst (e.g. Mediterranean coast and mountain) areas, stone construction is dominant – and in 2018 the art of dry-stone walling in countries such as France, Greece, Italy, Croatia and Spain was acknowledged by UNESCO to represent an intangible contribution to the cultural heritage of humanity.

A number of properties commonly found in indigenous architecture can be instructive for the future implementation of NBS (see overview of traditional materials in Table 1):

- Material is sourced from local vegetation or geological deposits (Heringer 2012).
- Reliance on non-renewable energy resources is low (Heringer 2012).
- Building form is thermally and structurally adapted to climate (Pearlmutter 2007).
- Durability is enhanced by regular maintenance (Georgi-Thomas & Zeumer 2012).
- Buildings are simple in construction and the materials are recyclable and compatible with biological cycles (Sauer 2012).

Building greening systems that are integrated with user lifestyles

Vegetation has always played an important role in the direct surroundings of houses. Living green plants symbolize life, health and prosperity – and have provided countless advantages to the inhabitants.

Trees planted directly next to the house are a regular feature of settlements in Europe (Wieland 1983). In central Europe, common species include Oak (*Quercus* spp.), Lime (*Tilia platyphyllos*), Ash (*Fraxinus excelsior*), Pear (*Pyrus* spp.) and Apple (*Malus domestica*). In southern Europe *Magnolia spec.*, *Acer* spp., *Olea europaea*, and Palm trees (*Palmae*) have been common, as has *Platanus* in western Europe. Trees have been appreciated for their aesthetic value as they indicate the seasons by flowers and colorful foliage. They also structure the direct surroundings of houses through shadow and light patterns, and deciduous trees selectively cast shadow in the summer and let the light pass during winter. Different species provide fruit, pharmaceuticals, fodder, and valuable wood, and attract insects for pollination. Trees have often carried a high cultural and mystic-religious importance as well. Nowadays, in spectacular cases such as the 'Hundertwasser Haus' in Vienna (Austria) or the 'Bosco Verticale' in Milano (Italy) trees are brought into and onto houses (Figure 2). At the same time, the appreciation for ordinary, traditional house trees and street trees is decreasing, as dwellers think they have less need for the trees' services. The work in gardens and the surroundings of houses has been outsourced (like in the Bosco Verticale) and for municipalities, maintenance costs and safety seem to be most important aspects (O'Sullivan *et al.* 2017). For instance, in Berlin, the number of street trees planted yearly decreased from over 8,000 in 1992 to just over 2,000 in 2018 – while the number of trees cut varied around a mean of around 4,800 per year (SenUVK Berlin 2019). This development, together with pests and diseases like ash dieback, Dutch elm disease (Jernelöv 2017), or the *Ceratocystis platani* and Massaria disease of *Platanus* species, are decreasing the number of house and street trees across Europe (Schmitt *et al.* 2014).

doi: 10.2166/bgs.2019.928

Table 1 | Traditional building materials and their relevant properties for implementation of NBS

Source	Material	Uses	Recyclable/reusable	Renewable	Environmental impact
Animals	wool/hair	blankets, carpets, textiles for insulation and shading	can be reused/reshaped or be used as an additive	yes (grows on animals)	CO_2 – impact of animal production (but could be a side product of food production)
		additive in clay or concrete	not reusable because bound in material		
	leather/pelt	tipi/yurt coverage	depending on process of dressing	yes	
	dung	floor, ground material	can be composted or burned	yes (animal waste product)	
		additive in clay or concrete	once dried out not easy to reuse, but can be returned to soil		
Plants	living plant material	sod for roofing	can be composted or burned at end of life	yes	binding of CO_2, habitat function
		climbers/clingers/creepers for bridges & sunscreens			
	wood	large-dimension timber & poles for primary structure	can be reused/reshaped repeatedly if not chemically treated, and composted or burned at end of life	yes, but careful management needed	binding of CO_2 during growth phase/harvesting, transport and production process creates positive or negative energy balance
		boards and planks			
		scantling, smaller pieces, shingles		yes	
	cork/bark	insulation material, flooring	can be reused or recycled	yes, but careful management needed	
	bamboo	roof structure, walls, fences, decoration	yes, but usually weak after initial use, can be composted or burned	yes, rapidly	
	rush, straw, thatch	roofs & ropes	can be composted	yes, rapidly	
	hemp	thermal insulation	yes	yes	
	paper	walls and screens	can be reused or recycled	yes	water/energy demand depending on production process, various chemicals added
		wallpaper	no (sticky glue on it)	partly	
Earth	water	ice buildings	can be reused or recycled	no	low
		component in clay or concrete	no		moderate?
	mud/clay	walls, floors, stoves	once dried out not easy to reuse, but can be returned to soil		low
		bricks	after firing not easily reusable, but can be returned to soil		average, depending on energy resource for firing
		tiles (glazed material)	reusable but not easily recyclable		depending on energy resource for firing and glazing
	stone	plaster	can be reused, recycled as gravel		depending on mining process and transport
		walls	can be reused, recycled as gravel		
		gravel	can be recycled in concrete		
	sand	glass	can be recycled		depending on mining process, transport and energy resource for production
	metal	various products	can be reused or recycled		
	lime	paint	can be returned to soil with other products		
	opus caementicium (lime and sand)	walls, floors, roofs	can be recycled in concrete		

Figure 2 | Examples of building integrated trees. Left: gardeners in front of Bosco Verticale in Milano, Italy and right: the Hundertwasser Haus in Vienna, Austria (photos: Thomas Nehls).

Façade greening is the melding of the artificial constructed and the naturally grown. There are very old examples for climbers and vines which were used to green buildings like the legendary hanging gardens of Semiramis and the shade creating wine-pergolas of Roman villas (Gothein 1926). Useful climbers and vines might have always been brought near houses and might have greened buildings and walls. *Hedera helix* is one of the most prominent examples, described in antiquity by Hippocrates of Kos (460–370 BC) and Pliny the Elder (23–79 AD) and in historical times as well (Turner 1538). In central Europe it has been used as a garden plant at least from the middle of the 16th century, and has been appreciated for its pharmaceutical potential. In the 18th century, the German polymath Goethe introduced climbers (vine stocks and ivy) to the upper class as part of garden and building art. There is a long experience with indirect façade greening using trellises or pergolas to support vines (e.g. *Vitis vinifera*), and with direct façade greening using plants that have adventitious roots like *Hedera helix* or climbing roots with adhesive pads such as *Parthenocissus tricuspedata*. While vines and climbers have long been associated with building damage and attraction of pests, a detailed study (Schlöesser 2003) showed that most inhabitants do have a positive attitude toward façade greening. It remains unclear why façade greening, given its aesthetical value, ecological importance and positive effects on mental and physical health (biophilia), is not more widespread in Europe. One initiative to remedy this is the I-BEST project currently in development by the University of Calabria, Italy: an innovative vegetated module for a green wall, which, by the integration with a rainwater harvesting system, allows for optimal urban water management and provides environmental and thermal benefits from a building scale to a city scale.

Green roofs are as old as façade greening, dating back to the Gardens of Semiramis as well. Traditionally, green roofs took the form of either sloped roofs covered with grass sod, or more sophisticated and intensively used roof gardens (Shafique *et al.* 2018). Turf roofs constructed for their insulation value have been a central feature of vernacular architecture in Iceland (van Hoof & van Dijken 2008), Greenland and Scandinavia (Ahrendt 2007). Roof gardens were constructed in ancient Rome to celebrate luxury, in the Renaissance to demonstrate the humanistic ideas of the owners, in the Baroque to symbolize the artificial look of the garden, and in classicism the antique examples flourished again (Ahrendt 2007). During all these epochs, the roofing technology changed only slightly, with bitumen, lead and copper as the main sealing materials. Apart from the turf roof, the motivations for hanging gardens have been the need for more productive land in dense cities, and the prestige to own a roof garden which symbolized wealth and power. With the advent of reinforced concrete it became much easier to construct flat roofs, with Hennebique among the first to recognize and to put into practice the greening of flat concrete roofs in 1901 (Kind-Barkauskas *et al.* 2001).

Recently, there is an increasing interest in developing more cost-effective green roof design, to use alternative building materials for liners and substrates, to combine greened roofs with solar energy production, and to create multi-purpose recreational space (Figure 3). Several cities are following this trend, and the EU Research and Innovation policy agenda promotes re-naturing cities and territorial resilience for socially and environmentally responsible communities, through the integration of NBS (EU 2015). However, a city's vision for the promotion and use of green roofs varies with the particularities of each place. Given the sustainability of green roofs over their full life cycle, policies are needed that encourage their use through regulation or financial incentives (such as water or property fee reduction), conditional to a pre-defined sustainable development goal attainment. Setting up quality standards for green roofs is important to scale up this NBS, with the German and the Spanish green roof guidelines being good examples for such (NTJ 11C 2012; FLL 2018).

Figure 3 | Example of cities sharing space with vegetation: Green roof at Praça de Lisboa, Porto, Portugal (photo: Cristina Calheiros).

Larger scale green topics in cities

Historic examples of NBS in city infrastructure are represented through food security systems in cities. There is a long culture of urban allotment gardens in European cities (Bell *et al.* 2016), as well as a history of guidelines that promote food production – for example in Athens, Greece. Here, in the development of the city after the 1960s there were guidelines for front-yards that allowed for the planting of two-three citrus trees, herb gardens and a small vegetable garden. Other examples include cultural landscape sites such as monasteries and cloisters in Europe, where herb gardens, vegetable gardens, orchards, vineyards and seed conservation banks were established and have contributed to food security in European towns for centuries, but also the establishment of cultural landscapes that protect the sloping banks of water reservoirs from erosion. The role of religious communities in protecting know-how regarding crop modification and soil fertility management practices is significant in Europe, but also around the world. Ancient precedents for integrated aquaculture include the chinampas of Mexico and the integrated rice paddy systems across parts of Asia, while what is today defined as aquaponics may be traced to the lowland Maya, followed by the Aztecs, who raised plants on rafts on the surface of a lake in approximately 1000 AD. Polyculture farming systems are common in far eastern countries as well (http://www.fao.org/3/y5098e/y5098e06.htm), as illustrated by the supplementary feeding of carp polycultures fertilized with duck manure in Hong Kong. Another example of historic NBS in infrastructure are the living-root bridges in the northeast Indian state of Meghalaya (Lewin 2012), constructed using the *Ficus elastica* tree and inspiring contemporary architects for techniques called 'baubotanik' (Ludwig 2012). In a similar way, practices of river restoration and bank stabilisation of soil-bioengineering that have been implemented in the rivers and streams of the Alps in Switzerland and Austria for years are being expanded in the design of contemporary ecological rivers and streams in cities.

Bioretention systems, biofilters, raingardens as well as constructed wetlands are green infrastructure systems implemented on the ground level for infiltration of water from different sources such as direct precipitation, runoff water, and in special cases polluted water from combined sewer overflow (CSO). These systems are areas that are excavated and filled with specific media depending on the actual multipurpose functions addressed, which may include maximizing the local infiltration capacity, thereby reducing stormwater volumes and relief pressure from the sewer system, serving as a stand-alone drainage system, supporting vegetation growth and providing sufficient water (Roy-Poirier *et al.* 2010). The last function is of utmost importance, as GI can only provide its benefits when there is sufficient availability of water. An important aspect is thereby related to the pollution of the runoff water. Heavy metals (HM) from roof tops and street runoff can degrade the groundwater quality. To cope with this, filtration media are developed to serve as barriers and retain pollutants such as HM (Haile & Fuerhacker 2018).

CURRENT PRACTICE: CIRCULAR SOLUTION CASE STUDIES

In the following, we present a series of case studies that demonstrate innovative approaches to the development and implementation of NBS in the built environment. The first is a research project on the development of alternative green building materials, and the second is a pilot project demonstrating a novel approach to edible green walls. The third is a multi-faceted experimental park combining a green roof demonstration with a number of other NBS technologies, and the final case study is a built example of a green building site in which the focus is on water-sensitive urban design (WSUD). The tables in the Appendix provide extensive lists of green projects in particular countries, illustrating the scope of implementation of different types of NBS.

Development of biocomposite building materials in the Negev, Israel

Project title: 'Biocomposite Building Materials Based on Agricultural Waste' (Scientific Team: David Pearlmutter, Erez Gal, Yaakov Florentin, Shahar Oannou, Francesca Ugolini)
Location: Sede Boker campus of Ben Gurion University of the Negev, Israel (30°48′N, 34°48′E)
Years of design and construction: 2016–2018
Nature-based solutions services: green building materials, recycled agricultural waste

Biocomposite building materials can have a significantly lower environmental footprint than conventional lightweight concrete, as they incorporate plant-based lightweight aggregates in a protective matrix – and these plants can sequester significant amounts of carbon in their growing phase.

doi: 10.2166/bgs.2019.928

By exploiting agricultural by-products to produce these alternative insulation materials, a further contribution is made towards reducing waste in a circular economy. At the same time, because these plant materials are concealed within an inert binder, they are not exposed to damage from fire, pests or rotting.

Experimental studies in the arid Negev Highlands of southern Israel were conducted to develop and test innovative biocomposite materials incorporating hemp shives and dried orange peels, respectively, as the insulating aggregate. As part of an ongoing life-cycle assessment, the thermal properties of each biocomposite were analyzed through laboratory testing and experiments using small test cells (1 × 1 × 0.6 m) with 20 cm thick walls.

Functionally graded hemp-lime biocomposite

Hemp-lime (HL) biocomposites are typically non-load bearing wall materials made from the woody core of the hemp plant (a non-psychoactive variety of *Cannabis sativa*), which is dried, cut into 'shives' and mixed with a lime binder. These lightweight insulating materials are non-toxic, reusable/recyclable and meet European fire and acoustic standards. Moreover, their production can be carbon 'negative', as CO_2 sequestered during the hemp plant's growth outweighs the net release of carbon in the production of lime (which gradually reabsorbs CO_2 through carbonation). In addition, HL is considered a 'breathable' material characterized by moisture buffering and improved indoor air quality. While the outer fibers of the hemp stalk are used for textiles and other products, its woody core comprises about 70% of the plant's weight and is usually treated as agricultural waste (Ip & Miller 2012; Zampori *et al.* 2013; Florentin *et al.* 2017).

In the first phase of the study, a life-cycle assessment of a homogenous HL biocomposite (450 kg m^{-3}) showed that the net carbon emissions of HL production are drastically reduced relative to a conventional building material with similar density and thermal properties, and that the magnitude of this reduction in embodied carbon is equivalent to about five years' worth of carbon emitted due to seasonal heating and cooling of a typical building (Florentin *et al.* 2017). The HL was also compared to a material combining internal thermal mass and external thermal insulation, which was better able to moderate the temperature fluctuations of a desert climate. However, because this variable density insulated concrete is based on cement and expanded polystyrene, it is very high in embodied energy and carbon. Thus the second phase of the study focused on the development and analysis of a variable density 'Gradient Hemp-Lime' block (Figure 4) which could significantly reduce operational as well as embodied energy and carbon emissions.

Figure 4 | (left) Gradient HL cross section; and (right) gradient HL test cell before plastering and roof installation, with homogenous hemp-lime cast on the corners to prevent thermal bridges. Photos: Y. Florentin.

The optimal configurations for the heavy and light phases of the gradient block were identified through laboratory testing, which included measurements of the density, thermal conductivity and volumetric heat capacity, and compressive strength of a wide range of samples in which various hemp-lime-sand-water ratios were cast and vibrated to separate the layers and attain a density gradient: from heavy on the bottom to light on top. A test cell was built of identical blocks based on the preferred ratio and vibration time, with the heavy side inward.

Results comparing the thermal performance of the gradient HL test cell to that of conventional lightweight concrete (Figure 5) show that during the early fall, the gradient hempcrete preforms better in terms of moderating indoor temperature fluctuations – maintaining a maximum temperature of 23 °C, which is in the range of thermal comfort. Thus the newly developed variable density hemp-lime composite block is seen in a preliminary analysis to offer superior thermal performance when compared with conventional materials of the same thickness. Furthermore, the 'gradient hemp-lime' shows considerable potential to substantially reduce the environmental impact of non-load bearing wall materials, especially in terms of CO_2 emissions, due to the carbon-negative bio-based aggregate made from the woody core of the hemp plant. Ongoing work, including a complete life cycle energy and carbon analysis, will provide further evidence of its potential as a sustainable material for the building industry.

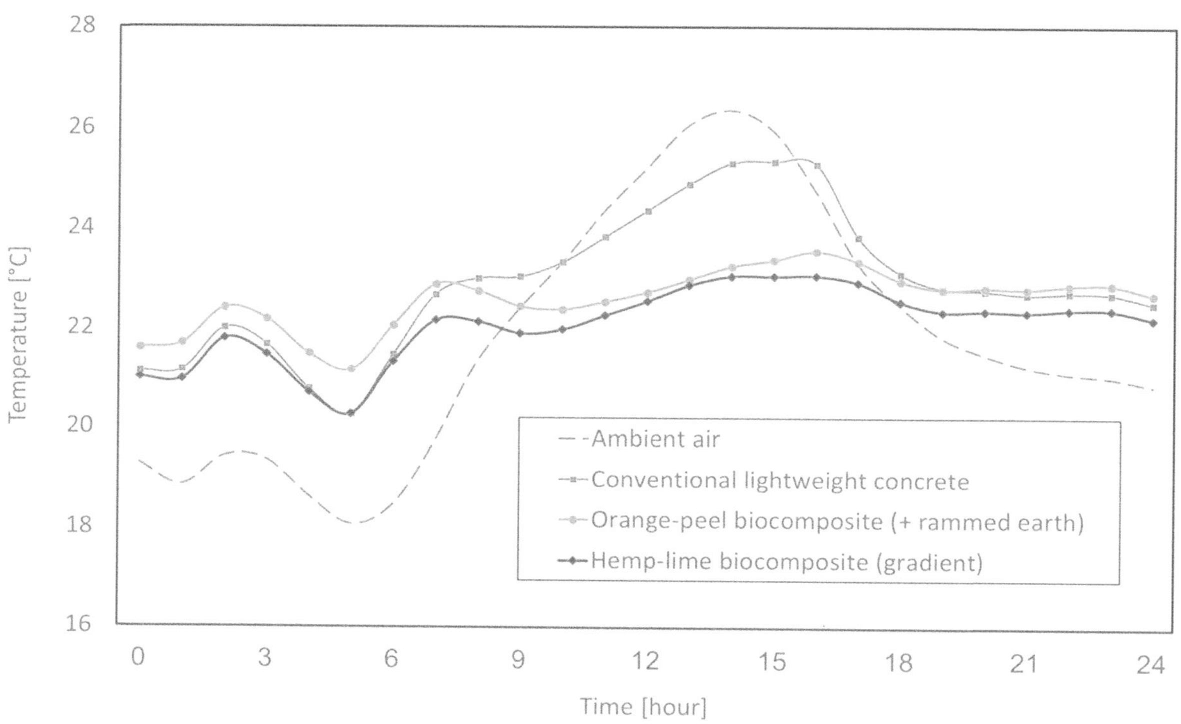

Figure 5 | Internal air temperature of test cells with wall construction based on the gradient hemp-lime and orange-peel biocomposites, compared with conventional lightweight concrete (18 October 2018 at the Sede Boker campus).

Citrus waste bio-composite and rammed earth

An innovative bio-composite building material based on agricultural waste was developed using dried orange peel (OP) particles as a lightweight aggregate and clay as a binder. This insulating biocomposite was integrated in a double-layer wall system, with the inner layer consisting of rammed earth (RE) for thermal mass.

doi: 10.2166/bgs.2019.928

The potential of this green material stems from the fact that most industrial waste from citrus (estimated at 15 million tons per year globally) is created in the juice production process and much of it consists of orange peels (Marín *et al.* 2007; USDA 2014). While this waste material may be used to produce animal feed or natural fertilizer for agriculture (Beccali *et al.* 2009), it is clear that orange peels represent a sizable potential resource for new applications such as building materials. The fact that much of the resource is already concentrated in processing plants for juice production means that its collection could be far more efficient than for other types of agricultural waste. The second wall material, rammed earth, is based on a traditional method of construction with naturally low embodied energy (Venkatarama Reddy & Prasanna Kumar 2010).

In the initial phase of the study, samples of the OP bio-composite with different proportions of OP and clay binder were compared in order to identify the highest proportion of insulating OP that could be used without compromising the mechanical stability of the composite as an infill (non-load bearing) material (Figure 6). The optimal configuration (with a density of 900 kg/m^3) was found to contain 46% OP and 52% clay by weight, with the small remainder consisting of Natural Hydraulic Lime (NHL) to minimize shrinkage. This proportion of insulating plant material is in agreement with general recommendations for hemp-lime mixtures as well. OP particle size fractions (from <1.2 to >4.7 mm) and water content were also optimized.

Figure 6 | (left) OP biocomposite samples with varying OP particle size, water content, and binder ratio; (right) test cell walls, with respective 10 cm-thick layers of external orange peel (OP) biocomposite insulation and internal rammed earth (RE) mass, before plastering and roof installation.

Test cell measurements (see Figure 5 above) showed that the orange-peel-rammed earth wall is able to moderate internal temperatures to a greater extent than lightweight concrete, and to nearly the same extent as the graded hemp-lime biocomposite. This indicates that the combination of external biocomposite insulation and internal thermal mass offers pronounced benefits for thermal comfort and energy savings.

Using an Element Analyzer to estimate the amount of carbon sequestered in the plant material, it was found that the carbon concentration (fraction of overall weight) in orange peels is 44%, which is virtually identical to that of hemp shives and similar to that of constituent organic molecules (46% for pectin and 49% for cellulose). This finding is being used for an ongoing life-cycle assessment of OP and HL biocomposites, with indications that both show potential as promising 'circular solutions' for the built environment.

Pilot demonstration of an edible green wall in Malmö, Sweden

Project title: Seved Edible green wall

Location: Malmö, Sweden (55°35′N 13°00′E)
Year of design and construction: 2013
Design: Odla i stan/Odlinsnätverket Seved/student från Malmö Högskola using system Gro-Wall
Contractor, installation: Peab (Nordic construction and civil engineering company) together with Peabskolan (a secondary school that trains within the framework of the national high school program building and construction)
Initiator: Föreningen Odlingsnätverket Seved (non-profit organization)
Co-financed: City of Malmo and MKB (*Malmö* Kommunala Bostads), Malmö Planterings- och försköningsförening (Planting and Beautification Association)
Nature-based solutions services: green edible wall, recycled material
Case study by: Alisa Korolova

The Seved Edible green wall in Malmo, Sweden was created in 2013 as a pilot project for the demonstration of vertical community gardens (Figure 7). The main idea was to inspire property owners to use the city space in a new way, as in many areas a lack of space does not allow residents to create community gardens or to grow plants in containers. The wall structure, which has a total area of 50 m^2 and a maximum height of 5 m, accommodates edible plants throughout the whole year – with specific plant types replaced depending on the season. The 'summer wall,' which includes plants such as strawberries, chard, lettuce, celery, spinach and herbs (oregano, lavender or rosemary), is cultivated from May until November. In November the 'summer wall' is replaced by the 'winter wall,' which is represented mostly by green cabbages and herbs like oregano and thyme. The selection is mainly determined by the visual aesthetics of the plants, with a preference for those that are bushy and compact – while plants that are especially sensitive to wind are generally avoided.

Figure 7 | Seved Edible garden (left); and Seved Edible green wall (right) (photos: Alisa Korolova).

The wall is made up of a modular system called Gro-Wall (https://www.gro-wall.com.au/). The system comprises a grid of compartments that are made from felt cloth bags, supplied by Mardam Agentur. The framework for attachment of the bags to the wall is made by Peab, who set up the wall together with the Peabskolan students. The irrigation is handled by an automated drip irrigation system, which is regulated via computer, together with manual watering by a management team from Odlingsnätverket Seved and the local community.

The project contributes to the city of Malmö and its circular economy through multiple ecosystem services: the increase of biodiversity, the safeguarding of edible planting processes, the nurturing of

doi: 10.2166/bgs.2019.928

local residents' relationship to natural food production, the local use and reuse of water that would otherwise be routed to the sewer, and the direct availability of vegetables on the dinner table of local families. It is a demonstration of nature-based solutions at a small scale, with innovative landscape elements designed to enhance circularity in terms of nutrients, food production and water reuse in available open space. More importantly, as a demonstration project it can serve as an example of what is possible for the whole city. In 2014, in fact, Malmö adopted a comprehensive urban farming program to get a better overview of the possibilities and with the aim of further promoting the city gardening concept.

Lessons learned:

- The Gro-Wall system is made of 100% recycled plastic, demonstrating the potential of circular solutions.
- The modular system allows diverse design, and can accommodate the use of vertical space for edible plants all year round.
- Cooperation between land/building owners, professionals and students, together with the engagement of a non-governmental organization (NGO), is a key to successful implementation.
- The project embodies important steps towards social sustainability, by providing support to the local community.

Demonstration of experimental green roof and allied technologies in the 'Urban Hydraulic Park'

Project title: 'Integrated and Sustainable management service for water-energy cycle in urban drainage systems' (Scientific Leader: Patrizia Piro)
Location: Rende, Italy (39°22′N 16°13′E)
Years of design and construction: 2011–2014
Area: The total area of all Experimental Sites located at the Urban Hydraulic Park is more than 700 m^2
Design: Urban Hydraulics and Hydrology Laboratory – University of Calabria (Scientific Supervisor Prof. Patrizia Piro) with the cooperation of the companies involved in the Project
Co-financed: Italian National Operative Project (PON) – PON01_02543
Nature-based solutions services: Green roof, Permeable Pavement, Stormwater Filter
Case study by: Patrizia Piro

The 'Urban Hydraulic Park' is an experimental demonstration site located at the University of Calabria, Italy, and specifically in the Vermicelli catchment (27.80 ha) where a series of Nature Based Solutions have been implemented to investigate their efficiency in terms of hydraulic, thermal and environmental benefits. The Park includes a green roof with a rainwater harvesting system, a permeable pavement, a stormwater filter, and a traditional sedimentation tank connected to a treatment unit. It is also equipped with a complex monitoring and acquisition system for the collection of climatic, hydrological, hydraulic, and thermo-physical data in real time.

The green roof experimental site shown in Figure 8 (Piro *et al.* 2019a) is parcelled into four hydraulically independent sectors with an area of about 40–50 m^2 each: three vegetated roofs and one conventional roof used as reference for the hydraulic and energetic analysis. While the three green roofs differ in terms of their drainage layers and/or the presence of vegetation, they all generally consist (from top to bottom) of a vegetated layer, a soil substrate, a permeable geotextile, a drainage and storage layer, an anti-root layer, and a waterproof membrane. Two sectors were covered by the same native Mediterranean plants (*Carpobrotus edulis*, *Dianthus gratianopolitanus*, and *Cerastium tomentosum*), while the third one hosts colonized plants. The water supply is guaranteed by reusing the green roof's outflow, which is collected in a storage tank (1.5 m^3) placed at the base of the building and distributed through a drip irrigation system during drought periods.

Figure 8 | The green roof experimental site at University of Calabria.

As documented in Palermo *et al.* (2019), the green roof described here demonstrated good hydraulic performance in terms of stormwater retention in a Mediterranean climate. By analysing the monitored data collected between October 2015 and September 2016 (62 rainfall events) at an event scale for one of the vegetated sectors, the mean value of subsurface runoff was found to be 32.0% – while the subsurface runoff coefficient, obtained considering stormwater events with precipitation depth of more than 8.0 mm (35 rainfall events), was 50.4%. In addition, the integration of the green roof and rainwater harvesting system implemented in the experimental site provided considerable benefits in terms of rational management of water resources. In this vein, a possible smart optimization of the green roof and rainwater harvesting system was described in Piro *et al.* (2019b).

The permeable pavement experimental site (Figure 9) was built in a portion of an existing car park. It has an area of around 380 m^2, divided into two sections: one of about 150 m^2 with permeable

doi: 10.2166/bgs.2019.928

Figure 9 | The permeable pavement experimental site at University of Calabria.

pavement, and the other left impermeable for use as a reference. It has an average slope of 2%, and a total profile depth of 0.98 m. The surface wear layer consists of porous concrete blocks characterized by high permeability (8 cm depth); while the base layer (35 cm depth); sub-base layer (45 cm depth) and bedding layer (5 cm depth) were defined by following the suggestions of the Interlocking Concrete Pavement Institute (ICPI), which recommends certain ASTM stone gradations. In order to evaluate the hydraulic behaviour of the permeable pavement, a reservoir element model was developed in Turco *et al.* (2018). In this study, the model was calibrated and validated against measured runoff collected on the specific experimental site described here.

Finally, the stormwater filter experimental site (Figure 10), installed downstream from the permeable pavement, is used to treat stormwater runoff discharged from the adjoining impervious parking lot. It has a surface area of around 125 m^2, an average slope of 2%, and a total profile depth of 0.75 m, and is covered by a soil substrate vegetated with Mediterranean species. A high permeability geotextile is placed between the soil substrate and the filter layer to prevent fine particles from migrating into the underlying layer. The filter layer is composed of highly permeable gravelly material. Finally, an impervious membrane at the bottom of the profile prevents water percolation into deeper horizons. Based on data collected at the experimental site (Brunetti *et al.* 2017), the

Figure 10 | The stormwater filter experimental site at University of Calabria.

unsaturated hydraulic properties of the stormwater filter were evaluated and the benefit of surrogate-based modelling was demonstrated in the numerical analysis of sustainable solutions.

Lessons learned:

- Each NBS represents a 'low impact development' solution, with a specific stratigraphy designed by taking into account the climate condition and the regulations in force.
- From analysis of the monitoring data, all of these solutions improve urban stormwater management in terms of surface runoff mitigation and water quality enhancement.
- From investigation of energetic data, the green roof proves to be suitable for reducing the temperature variation in the building and mitigating the urban heat island effect.
- The LCA analysis carried out for the specific green roof and permeable pavement confirms the sustainability of these low-impact infrastructures (Maiolo *et al.* 2017).

Photo credits: Urban Hydraulic and Hydrology Laboratory (http://www.giare.eu/) – Scientific Supervisor: Patrizia Piro

Zollhallen Platz in Freiburg, Germany

Project title: Zollhallen Platz
Location: Freiburg, Germany (48°0′ N/7°50′ E)
Year of design: 2009–2010
Year of construction: 2011
Area: 5,600 m^2
Design company: Ramboll Studio Dreiseitl
Nature-based solutions services: Water-sensitive urban design, biodiverse local planting, reuse of existing material in new design
Case study by: Dimitra Theochari

Zollhallen Plaza is located at the entrance to a historic train station designated for customs that was restored in 2009 (Figures 11–14). Although the scale of the plaza is small, and it was initially designated to be a simple hardscape area for the new users of this public building, the design team set an ambitious target: to disconnect the plaza from the sewer system, and to create a small-scale urban plaza that would be an example of water-sensitive urban design (WSUD). A series of infiltration points located in the plaza through planters, connected with subsurface gravel trenches and in-built filter medium, are used to reduce the hydraulic overload on the sewer system. The plaza is designed

Figure 11 | Local planting and view to the old train station of Freiburg. The details of the train tracks and the memory of the area is brought back in a system showing applicability of circular city material recycling applications (left), and details of seating designed as being peeled-off the ground to allow for maximizing the space for permeable green surfaces that are entry points to the infiltration system to the underground storage, but also to the groundwater reservoir (right).

doi: 10.2166/bgs.2019.928

Figure 12 | Connecting the hardscape to the green zones, extending the ecology and urban nature qualities of the site.

Figure 13 | Details of planting and the recycled train tracks from the site.

Figure 14 | View of the seating as used by visitors and locals. Area to relax and contemplate urban nature (left); and seating in the green zone and looking out to the city (right). The plaza stormwater system is disconnected from the municipality sewer.

to hold on its surface the volume of water generated by a 20-year, a 50-year and a 100-year rain-event, and to provide flood safety to the city while recharging the ground water table.

In terms of material reuse on site, the rail pieces from the rail yard are reused to structure the inlets of the infiltration with perennials and ornamental grasses, creating an aesthetic appeal of small scape colourful planting. All of the hardscape materials are high-quality demolition materials recycled from the old rail yard, a fact that makes this case study a true example of NBS in circular city principles. In this way, the design evokes the continuity of special historic and cultural features, and the memory of people who worked in the area – contributing further to the social dimension of ecosystem services.

This case study was designed and developed as a real-life project for the city of Freiburg, setting new targets for detaching the plaza from the sewer system and creating a sustainable water irrigation and groundwater infiltration system. Quantitative data on the functioning of the system are not available, but there is common experience from the site and monitoring of the construction quality – as is typical for any city project in the Baden-Württemberg in Germany – that attest to the high efficiency of the system. Qualitatively, the NBS derived from Zollhallen Plaza include an increase in the permeability of the plaza, which does not create a greater burden for the city's stormwater system, and which contributes to the strengthening of the biodiversity of the city of Freiburg through the planting selection. The water circulation and reuse for irrigation of the planting throughout project responds to considerations of NBS in the context of a Circular City.

Lessons learned:

- The hardscape of a small city plaza can be used to create a stormwater management system and extreme flooding management system, independent of the city sewer.
- The project demonstrates circular city applications for material and water that falls on site.
- An on-site water circulation system with an underground storage tank is charged with infiltration through permeable paving.

Photo credits:

Dimitra Theochari

CRITICAL ANALYSIS OF NBS IN THE BUILT ENVIRONMENT

Ecosystem services and disservices

Living vegetation that is integrated with or adjacent to buildings – including green roofs, balconies and façades, as well as gardens, parks and isolated trees – are part of a city's green infrastructure. Despite their urbanized surroundings, these green spaces host a wide range of animal species.

Whether the vegetation is long-established or newly planted, many vertebrates and invertebrates find shelter and food in these areas, including in the soil. Thus, the greening of the built environment contributes to biodiversity by decreasing the impacts of habitat fragmentation in urban ecosystems, and by increasing the permeability of the urban matrix (Martin-Queller *et al.* 2017). This biodiversity also supports a wide range of ecosystem services that contribute to the improvement of human health and wellbeing (Pinho *et al.* 2017).

Urban greening is invaluable for achieving the three main objectives of integrated water management – enhancing water availability, improving water quality and reducing water-related risks (WWAP 2018). Understory plant species contribute to regulating water quality by controlling nutrient runoff (Livesley *et al.* 2016), thus decreasing the resources required to deal with polluted water. Vegetation provides air quality regulation, by preventing pollutants from reaching buildings (Matos *et al.* 2019), thus decreasing the impacts and cost of air pollution in sensitive areas. Local invertebrates contribute to pollination, which together with birds also provide pest regulation (Mexia *et al.* 2018). Green spaces also increase the cultural sense of attachment to the place and to nature, with further benefits to wellbeing (Luz *et al.* 2019).

In addition to providing ecosystem services, however, the establishment and restoration of planted areas in the midst of buildings may increase the exposure of residents to allergenic material, mostly from vegetation pollen. A list of allergenicity of common tree species is available, and should be taken into account when building or restoring urban green spaces (Cariñanos *et al.* 2019). High vegetation density, although a bonus for biodiversity, may increase some people's sense of insecurity, and thus should be evaluated for each case. High moisture, promoted by excessive watering, may cause an increase in local pest populations.

However, balanced urban ecosystems – most notably those in which the use of insecticides is limited – are likely to mitigate this effect by increasing the local populations of insect parasites or predators, such as birds or other insects (Sanchez-Bayo & Wyckhuys 2019). Planting trees near air pollution sources will increase the local deposition of pollutants (Santos *et al.* 2017), screening them from the urban air to which urban residents are exposed. While this may benefit those located beyond the trees, however, it may also increase pollution loads for those located between the air pollution source and the trees. Although such trade-offs may have no measurable effect on the average air pollution status in the city, careful planning must take them into account in order to provide the highest air quality where it is needed most.

Other planning and management options also influence the ecosystem services provided by urban green infrastructure, because the same choice of habitat (e.g. grass vs forest) cannot maximize all ecosystem services. For example, native vegetation can provide many support and regulation ecosystem services, such air and microclimate regulation, due to a more complex vegetation structure (Vieira *et al.* 2018). However, other services such as pollination are not favoured by such dense vegetation (Mexia *et al.* 2018) due to the exclusion of some birds' functional types. Additionally, vegetation density is not related to people's willingness to visit parks, highlighting that cultural services are not necessarily increased in areas where support and regulation ecosystem services are the highest (Shanahan *et al.* 2015).

Implementation of NBS for a resourceful circular city: the role of EU policy and international policy drivers

Through its general environmental legislation, the EU has policies in place to promote nature-based solutions in European cities. These policies aim to ensure that urban residents can enjoy clean air and water and avoid excessive exposure to noise, and that cities deal properly with waste, protect nature and biodiversity, and promote green infrastructure. The European Green Capital (http://ec.europa.eu/environment/europeangreencapital/index_en.htm) and European Green Leaf (http://ec.europa.eu/environment/europeangreencapital/europeangreenleaf/index.html) Award programs are tangible initiatives which allow cities to showcase their environmental performance, and policy instruments such as the Circular Economy Action Plan (http://ec.europa.eu/environment/circular-economy/index_en.htm) even address issues such as resource efficiency and raw materials – which can find application in construction, as well as other sectors.

The EU has also elaborated a vision on how to build a sustainable finance strategy (https://ec.europa.eu/info/business-economy-euro/banking-and-finance/sustainable-finance_en) that can cope with the economic challenges of implementing NBS. With an emphasis on funding society's long-term needs, the goal is to strengthen financial stability by incorporating environmental, social and governance (ESG) factors into investment decision-making. Tangible actions include establishing an EU classification system for sustainable activities, creating standards and labels for green financial products, and fostering investment in sustainable projects.

By adopting the Paris Agreement on climate change and the UN 2030 Agenda for Sustainable Development in 2015, governments from Europe and around the world agreed on a more sustainable and resilient path for our planet, providing environmental, social and economic benefits for all. Better alignment of their agendas on sustainable development could unlock the mechanisms needed to achieve a coordinated perspective, utilizing NBS as a cost-effective way of creating greener, more sustainable and more competitive circular economies.

However, the EU still needs a tighter focus on nature-based solutions that are targeted directly to green building materials, systems and sites. The main legislative instrument dedicated specifically to buildings is the 2010 Energy Performance of Buildings Directive (EPBD), which like the 2012 Energy Efficiency Directive focuses exclusively on operational energy – and as such, does not cover the overall life cycle impacts of buildings that are crucial to any vision of a circular economy. The intent of achieving 'circular buildings' is to comprehensively reduce life cycle impacts, and at the same time provide healthy and comfortable spaces for urban dwellers.

This sort of holistic approach to circularity is still lacking in existing certification systems, though minimum requirements for circular design are in fact part of an EU 2017 framework called Level(s), whose development is ongoing.

The lack of integration between NBS and building-related policy is illustrated by an emphasis on design strategies and products which save energy in a building's operation but have high levels of embodied energy and carbon in their production. For example, the shading of windows to prevent overheating is an operationally energy-efficient strategy, but one which is typically implemented using elements such as roller shutters made from materials like aluminium or other metals and plastics which are energy-intensive in their production. Alternatively, systems that use live plants for shading can reduce the building's environmental footprint in a more holistic manner – but they have yet to be mainstreamed in practice or prioritized in policy. In general, the greening of built surfaces – through the use of plant-based building materials, green roofs and walls, and water-sensitive site design – is still not sufficiently promoted.

CONCLUSIONS: ADVANCING THE IMPLEMENTATION OF NBS IN THE BUILT ENVIRONMENT

In light of the preceding discussion, we may ask why there are no specific EU policies, whether in the form of regulations, incentives, or educational tools, that explicitly promote NBS in the built urban environment. One question to be addressed regards the legal obstacles: once appropriate green technologies are available, specific policies need to be formulated in order to provide a regulatory framework that will allow these alternative approaches to find application in the construction sector, which is very strictly regulated (https://www.cen.eu/Pages/default.aspx). Part of this policy formulation involves the definition not just of technical measures, but of long-term criteria and aims that are aligned with the transition to a circular economy.

doi: 10.2166/bgs.2019.928

If the problem is a lack of information and precedents, then the review of case studies presented above may be seen as a step forward in providing such information. Overall, these case studies demonstrate that existing and developing nature-based solutions in the built environment are indeed effective and promising. We would argue that it is the implementation of these existing approaches that is firstly lacking, and this alone would hold great benefits in terms of sustainable and resilient development. In addition, further research is needed into the technical and societal aspects of these solutions to make them more powerful and appropriate for our future cities.

To the extent that a limiting factor is represented by a lack of technical data and adeptness with appropriate tools, there are existing sustainability indicators that can be leveraged. These include Life-Cycle Assessment (LCA), Material Flow Analysis (MFA), and many other metrics that address circular flow of energy, water and materials in a life cycle perspective. Undoubtedly there is still a need for R&D on design tools that are scientifically robust and at the same time accessible to designers and consultants. Among other things, better definitions of minimum performance levels, multiple functions, and valorisation of side effects should be included in decision making systems regarding the implementation of NBS in the built environment.

Further research into building-related NBS thus needs to be prioritized. The indicative case studies presented here suggest that funding for further pilot studies, demonstrations, and experimental monitoring data is crucial. Systematic analysis of the performance of NBS needs to be conducted with reference to conventional technical systems, so that the relative benefits may be quantified. Cost-benefit analysis of NBS against other solutions can only be done in a case-by-case analysis, but an emerging pattern when evaluating the success of NBS is that it must: (1) involve a large number of stakeholders and their interests; and (2) be shown to be a sustainable solution not only ecologically, but economical and socially as well (Nesshover *et al.* 2017).

We may also draw specific conclusion regarding the three levels of the built environment addressed in this paper: green building materials, systems and sites.

- *Regarding green building materials*: We emphasize that in contrast to the vernacular, modern construction is largely based on manufactured materials that offer reliability and convenience – but that these properties are usually attained through high-temperature, resource-intensive processes that generate considerable waste and atmospheric emissions. Biocomposite materials, which draw on the environmental benefits of nature-based ingredients but also fulfil the required attributes of modern construction thanks to their protective matrix, can make a significant contribution in the transition to a circular economy. A holistic life-cycle analysis can help to identify the sustainability potential and weak-points of such solutions.
- *Regarding green building systems*: A huge variety of building greening systems are already marketed, all promising to mitigate negative impacts of built environments. However, they come with different environmental, economic and social costs, onsite and offsite – 'green' does not mandatorily mean 'sustainable'. The integration of circular economy approaches can improve building greening systems' efficiency, sufficiency and consistency. Prior to investing tax money in NBS it will be crucial to prioritize these systems regarding their ecological, economic and social impacts using holistic life-cycle analyses. LCAs can be used to identify both optimization potentials and best practices. In order to conduct them in a reproducible and transparent way, a harmonised set of a-priori weighted assessment criteria has to be employed and clearly documented.
- *Regarding green building sites*: Especially at the scale of semi-public and public open spaces, the effective integration of NBS can meaningfully contribute to resolving larger scale societal challenges locally, by making cities more healthy and inclusive as well as more resource-efficient. In particular, urban food production can support action toward urban resilience and social cohesion. Water-sensitive urban design approaches and a shift from conventional 'grey' infrastructure to blue-green infrastructure solutions offer flood safety, resilience and opportunities to close the water cycle on every site. Moreover, green

building sites contribute in terms of noise mitigation, biodiversity connectivity, microclimate mitigation and outdoor comfort, while often their benefits are multiple and overlapping, including these and many more ecosystem services. It should be noted that the green building sites are also the designated areas where the offset of the carbon footprint of the development can be placed.

Finally, it is crucial to see green building materials, systems and sites as part of a holistic urban web. When different types of NBS are combined in an integral way – creating a multi-scale network of urban blue-green infrastructure – then the paradigm shift toward circularity will indeed be on its way.

ACKNOWLEDGEMENTS

The work is funded within the COST Action CA17133 Circular City ('Implementing nature based solutions for creating a resourceful circular city', http://www.circular-city.eu, duration 22 October 2018–21 October 2022). COST Actions are funded within the EU Horizon 2020 Programme. The authors are grateful for the support. Ramboll Studio Dreiseitl for the Zollhallen Plaza case study. The 'Urban Hydraulic Park' Demonstration site was funded by the Italian Operational Project (PON) – Research and Competitiveness for the convergence regions 2007/2013 — I Axis 'Support to structural changes' operative objective 4.1.1.1. 'Scientific-technological generators of transformation processes of the productive system and creation of new sectors' Action II: 'Interventions to support industrial research'. The 'I-Best' Project (Scientific Leader Prof. Mario Maiolo) is funded by Italian National Operational Program 'Enterprise and Competitiveness' 2014–2020 ERDF – I AXIS 'Innovation' - Action 1.1.3 – 'Support for the economic enhancement of innovation through experimentation and the adoption of innovative solutions in processes, products and organizational formulas, as well as through the financing of the industrialization of research results'.

SUPPLEMENTARY MATERIAL

The Supplementary Material for this paper is available at http://dx.doi.org/10.2166/bgs.2019.928.

REFERENCES

Ahrendt, J. 2007 *HISTORISCHE GRÜNDÄCHER: Ihr Entwicklungsgang bis zur Erfindung des Eisenbetons, Teil I. Dissertation* (PhD Thesis), Technische Universitaet Berlin, German. online: urn:nbn:de:kobv:83-opus-14771.

Beccali, M., Cellura, M., Iudicello, M. & Mistretta, M. 2009 Resource consumption and environmental impacts of the agrofood sector: life cycle assessment of Italian citrus-based products. *Environmental Management* **43** (4), 707–724.

Bell, S., Fox-kämper, R., Keshavarz, N., Benson, M. & Caputo, S. 2016 Urban Allotment Gardens in Europe – 9 March 2016.

Blaser, W. 1982 *Elementare Bauformen: Quellen moderner Architektur (Elemental Building Forms: Sources of Modern Architecture).* Beton-Verlag GmbH, Duesseldorf, Germany.

Brown, A. 2012 *Just Enough: Lessons in Living Green From Traditional Japan.* Tuttle Publishing, North Clarendon, VT, USA.

Browning, W. D., Ryan, C. O. & Clancy, J. O. 2014 *14 Patterns of Biophilic Design.* Terrapin Bright Green, LLC, New York.

Brunetti, G., Šimůnek, J., Turco, M. & Piro, P. 2017 On the use of surrogate-based modeling for the numerical analysis of low impact development techniques. *Journal of Hydrology* **548**, 263–277.

Buchin, O., Hoelscher, M., Nehls, T. & Zieglera, F. 2016 Evaluation of the health-risk reduction potential of countermeasures to urban heat islands. *Energy and Buildings* **114**, 27–37.

Cariñanos, P., Grilo, F., Pinho, P., Casares-Porcel, M., Branquinho, C., Acil, N., Andreucci, M. B., Anjos, A., Bianco, P. M., Brini, S., Calaza-Martínez, P., Calvo, E., Carrari, E., Castro, J., Chiesura, A., Correia, O., Gonçalves, A., Gonçalves, P., Mexia, T., Mirabile, M., Paoletti, E., Santos-Reis, M., Semenzato, P. & Vilhar, U. 2019 Estimation of the allergenic potential of urban trees and urban parks: towards the healthy design of urban green spaces of the future. *International Journal of Environmental Research and Public Health* **16** (8), 1357.

Elmqvist, T., Fragkias, M., Goodness, J., Güneralp, B., Marcotullio, P. J., McDonald, R. I., Parnell, S., Schewenius, M., Sendstad, M., Seto, K. C. & Wilkinson, C. (eds) 2013 *Urbanization, Biodiversity and Ecosystem Services: Challenges and*

Opportunities. A Global Assessment A Part of the Cities and Biodiversity Outlook Project. Springer, Dordrecht, Heidelberg, New York, London.

Enerdata 2012 *Energy Efficiency Trends in Buildings in the EU: Lessons From the ODYSSEE MURE Project*. https://energiatalgud.ee/img_auth.php/6/68/Enerdata._Energy_Efficiency_Trends_in_Buildings_in_the_EU._2012.pdf (accessed 30 May 2019).

European Union 2015 *Towards an EU Research and Innovation Policy Agenda for Nature-Based Solutions & Re-Naturing Cities Final Report of the Horizon 2020 Expert Group on Nature-Based Solutions and Re-Naturing Cities*.

FLL 2018 *Green Roof Guidelines – Guidelines for the Planning, Construction and Maintenance of Green Roofs*, 2018 edn. Forschungsgesellschaft Landschaftsentwicklung Landschaftsbau e.V. (FLL), Bonn.

Florentin, Y., Pearlmutter, D., Givoni, B. & Gal, E. 2017 A life-cycle energy and carbon analysis of hemp-lime biocomposite building materials. *Energy and Buildings* **156**, 293–305.

Georgi-Thomas, A. & Zeumer, M. 2012 Einfach nachhaltig (Simply Sustainable). In: *Im Detail: Einfach Bauen Zwei: nachhaltig, kostengünstig, lokal (In Detail: Simply Building Two: Sustainable, Economical, Local)* (Schittich, C., ed.). Institut für internationale Architektur-Dokumentation GmbH & Co. KG, Postfach 201054, D-80010, Muenchen, Germany, pp. 42–49.

Gothein, M. L. 1926 *Geschichte der Gartenkunst 1: Von Aegypten bis zur Renaissance in Italien, Spanien und Portugal (History of Garden Art 1: From Egypt to the Renaissance of Italy, Spain and Portugal)* (Diedrichs, J., ed.).

Haile, T. M. & Fuerhacker, M. 2018 Simultaneous adsorption of heavy metals from roadway stormwater runoff using different filter media in column studies. *Water* **10** (9), 160.

Heringer, A. 2012 Einfach lokal (Simply Local). In: *Im Detail: Einfach Bauen Zwei: nachhaltig, kostengünstig, lokal (In Detail: Simply Building Two: Sustainable, Economical, Local)* (Schittich, C., ed.). Institut für internationale Architektur-Dokumentation GmbH & Co. KG, Postfach 201054, D-80010, Muenchen, Germany, pp. 50–55.

Huberman, N., Pearlmutter, D., Gal, E. & Meir, I. A. 2015 Optimizing structural roof form for life-cycle energy efficiency. *Energy and Buildings* **104**, 336–349.

Ip, K. & Miller, A. 2012 Life cycle greenhouse gas emissions of hemp-lime wall constructions in the UK. *Resources, Conservation and Recycling* **69**, 1–9.

IUCN Commission on Ecosystem Management 2019 *Nature-based Solutions*. https://www.iucn.org/commissions/commission-ecosystem-management/our-work/nature-based-solutions (accessed 30 May 2019).

Jernelöv, A. 2017 Dutch Elm Disease in Europe and North America. In: *The Long-Term Fate of Invasive Species*. Springer, Cham, pp. 161–176.

Kabisch, N., Korn, H. & Stadler, J. (eds) 2017 *Nature-Based Solutions to Climate Change Adaptation in Urban Areas Linkages Between Science, Policy and Practice* (Bonn, A., ed.). Springer. doi:10.1007/978-3-319-56091-5.

Kind-Barkauskas, F., Kauhsen, B., Polónyi, S. & Brandt, J. 2001 *Beton Atlas*. Detail, Berlin, p. 298.

Langergraber, G., Pucher, B., Simperler, L., Kisser, J., Katsou, E., Buehler, D., Garcia Mateo, M. C. & Atanasova, N. 2019 *Implementing Nature-Based Solutions for Creating A Resourceful Circular City* (submitted). *Blue-Green Systems*.

Lehner, E. 2016 *Roots of Architecture: Building Traditions in Non-European Cultures*. IVA-ICRA Publishing, Institute for Comparative Architecture, Vienna, Austria.

Lewin, B. 2012 *India's Living Bridges*. Reader's Digest Australia, pp. 82–89. Archived from the original on 16 November 2012.

Livesley, S. J., Ossola, A., Threlfall, C. G., Hahs, A. K. & Williams, N. S. G. 2016 Soil carbon and carbon/nitrogen ratio change under tree canopy, tall grass, and turf grass areas of urban green space. *Journal of Environmental Quality* **45**, 215–223.

Ludwig, F. 2012 Baubotanik Tower. In: *Bio Design: Nature + Science + Creativity* (Myers, W., ed.). The Museum of Modern Art, New York, p. 304.

Luz, A., Buijs, M., Aleixo, C., Metelo, I., Grilo, F., Branquinho, C., Santos-Reis, M. & Pinho, P. 2019 Should I stay or should I go? modelling the fluxes of urban residents to visit green spaces. *Urban Forestry & Urban Greening* **40**, 195–203.

Maiolo, M., Carini, M., Capano, G. & Piro, P. 2017 Synthetic sustainability index (SSI) based on life cycle assessment approach of low impact development in the Mediterranean area. *Cogent Engineering* **4** (1), 1410272. https://doi.org/10.1080/23311916.2017.1410272.

Marín, F. R., Soler-Rivas, C., Benavente-García, O., Castillo, J. & Perez-Alvarez, J. A. 2007 By-products from different citrus processes as a source of customized functional fibres. *Food Chemistry* **100** (2), 736–741.

Martin-Queller, E., Albert, C. H., Dumas, P. J. & Saatkamp, A. 2017 Islands, mainland, and terrestrial fragments: how isolation shapes plant diversity. *Ecology and Evolution* **7**, 6904–6917.

Matos, P., Vieira, J., Rocha, B., Branquinho, C. & Pinho, P. 2019 Modeling the provision of air-quality regulation ecosystem service provided by urban green spaces using lichens as ecological indicators. *Science of the Total Environment* **665**, 521–530.

Mexia, T., Vieira, J., Principe, A., Anjos, A., Silva, P., Lopes, N., Freitas, C., Santos-Reis, M., Correia, O., Branquinho, C. & Pinho, P. 2018 Ecosystem services: urban parks under a magnifying glass. *Environmental Research* **160**, 469–478.

Nesshover, C., Assmuth, T., Irvine, K. N., Rusch, G. M., Waylen, K. A., Delbaere, B., Haase, D., Jones-Walters, L., Keune, H., Kovacs, E., Krauze, K., Külvik, M., Rey, F., Dijk, J., Vistad, O. I., Wilkinson, M. E. & Wittmer, H. 2017 The science, policy and practice of nature-based solutions: an interdisciplinary perspective. *Science of the Total Environment* **579**, 1215–1227.

NTJ 11C 2012 *Normas Técnológicas de jardinería y paisajismo - Ajardinamientos especiales: cubiertas verdes*. Fundació de la Jardineria i el Paisatge.

Olmsted, Frederick Law 1865 *Introduction to Yosemite and the Mariposa Grove: A Preliminary Report*.

O'Sullivan, O. S., Holt, A. R., Warren, P. H. & Evans, K. L. 2017 Optimising UK urban road verge contributions to biodiversity and ecosystem services with cost-effective management. *Journal of Environmental Management* **191**, 162–171.

Palermo, S. A., Turco, M., Principato, F. & Piro, P. 2019 Hydrological effectiveness of an extensive green roof in Mediterranean climate. *Water* **11** (7), 1378.

Pearlmutter, D. 2007 Architecture and climate: the environmental continuum. *Geography Compass* **1** (4), 752–778.

Pinho, P., Moretti, M., Luz, A. C., Grilo, F., Vieira, J., Luís, L., Rosalino, L. J., Martins-Loução, M. A., Santos-Reis, M., Correia, O., Garcia-Pereira, P., Gonçalves, P., Matos, P., de Carvalho, R. C., Rebelo, R., Dias, T., Mexia, T. & Branquinho, C. 2017 Biodiversity as Support for Ecosystem Services and Human Wellbeing. In: *The Urban Forest: Cultivating Green Infrastructure for People and the Environment* (Pearlmutter, D., Calfapietra, C., Samson, R., O'Brien, L., Krajter Ostoić, S., Sanesi, G. & Alonso del Amo, R., eds). Springer, Cham. Future City **7**, pp. 67–78.

Piro, P., Carbone, M., Morimanno, F. & Palermo, S. A. 2019a Simple flowmeter device for LID systems: from laboratory procedure to full-scale implementation. *Flow Measurement and Instrumentation* **65**, 240–249.

Piro, P., Turco, M., Palermo, S. A., Principato, F. & Brunetti, G. 2019b A Comprehensive Approach to Stormwater Management Problems in the Next Generation Drainage Networks. In: *The Internet of Things for Smart Urban Ecosystems* (Piro, P., ed.). Springer, Cham, pp. 275–304.

Prieto-Sandoval, V., Jaca, C. & Ormazabal, M. 2018 Towards a consensus on the circular economy. *Journal of Cleaner Production* **179**, 605–615.

Rodgers, P. 2019 *The Living Root Bridge Project: A Database for Tribal Botanical Architecture.* https://livingrootbridges.com/about/ (accessed 30 May 2019).

Roy-Poirier, A., Champagne, P. & Filion, Y. 2010 Review of bioretention system research and design: past, present, and future. *Journal of Environmental Engineering* **136**, 9.

Saaroni, H., Amorim, J. H., Hiemstra, J. A. & Pearlmutter, D. 2018 Urban green infrastructure as a tool for urban heat mitigation: survey of research methodologies and findings across different climatic regions. *Urban Climate* **24**, 94–110.

Sanchez-Bayo, F. & Wyckhuys, K. A. G. 2019 Worldwide decline of the entomofauna: a review of its drivers. *Biological Conservation* **232**, 8–27.

Santos, A., Pinho, P., Munzi, S., Botelho, M. J., Palma-Oliveira, J. M. & Branquinho, C. 2017 The role of forest in mitigating the impact of atmospheric dust pollution in a mixed landscape. *Environmental Science and Pollution Research* **24**, 12038–12048.

Sauer, C. 2012 Einfach konstruieren (Simply constructed). In: *Im Detail: Einfach Bauen Zwei: nachhaltig, kostengünstig, lokal (In Detail: Simple Building Two: Sustainable, Economical, Local)* (Schittich, C., ed.). Institut für internationale Architektur-Dokumentation GmbH & Co. KG, Postfach 201054, D-80010, Muenchen, Germany, pp. 12–23.

Schlösser, S. 2003 *Zurakzeptanz von fassadenbegrünungen. Meinungsbilderkölnerbürger – einebevölkerungsumfrage.* PhD, Univ. Köln, p. 184.

Schmitt, U., Luer, B., Dujesiefken, D. & Koch, G. 2014 The massaria disease of plane trees: its wood decay mechanism. *IAWA Journal* **35** (4), 395–406.

SenUVK 2019 *Senate Department for Environment, Transportation and Climate Protection. Bestandszahlen zu Strassenbämen.* https://www.berlin.de/senuvk/umwelt/stadtgruen/stadtbaeume/de/daten_fakten/ uebersichten/index.shtml (accessed 1 November 2019).

Shafique, M., Kim, R. & Rafiq, M. 2018 Green roof benefits, opportunities and challenges – A review. *Renewable and Sustainable Energy Reviews* **90**, 757–773.

Shanahan, D. F., Lin, B. B., Gaston, K. J., Bush, R. & Fuller, R. A. 2015 What is the role of trees and remnant vegetation in attracting people to urban parks? *Landscape Ecology* **30**, 153–165.

Turco, M., Brunetti, G., Carbone, M. & Piro, P. 2018 Modelling the hydraulic behaviour of permeable pavements through a reservoir element model. *International Multidisciplinary Scientific GeoConference on SGEM: Surveying Geology & Mining Ecology Management* **18**, 507–514.

Turner, W. 1538 *Libellus de re Herbaria Novus.* https://doi.org/10.5962/bhl.title.59358 (accessed 30 May 2019)

USDA 2014 *Citrus: World Markets and Trades.* United States Department of Agriculture. Retrieved from http://www.fas.usda.gov/data/citrus-world-markets-and-trade (accessed 30 May 2019).

US Energy Information Administration 2017 International Energy Outlook 2017. https://www.eia.gov/outlooks/archive/ieo17/ (accessed 30 May 2019).

van Hoof, J. & van Dijken, F. 2008 The historical turf farms of Iceland: architecture, building technology and the indoor environment. *Building and Environment* **43** (6), 1023–1030.

Venkatarama Reddy, B. V. & Prasanna Kumar, P. 2010 Embodied energy in cement stabilised rammed earth walls. *Energy and Buildings* **42**, 380–385.

Vieira, J., Matos, P., Mexia, T., Silva, P., Lopes, N., Freitas, C., Correia, O., Santos-Reis, M., Branquinho, C. & Pinho, P. 2018 Green spaces are not all the same for the provision of air purification and climate regulation services: the case of urban parks. *Environmental Research* **160**, 306–313.

Wieland, D. 1983 Topographie. Bauen und Bewahren: der Hausbaum. Bayrischer Rundfunk (German) https://www.br.de/mediathek/video/topographie-1983-der-hausbaum-av:5896bbf2ab0d0d0012fcd4f5 (accessed 30 May 2019).

WWAP (United Nations World Water Assessment Programme)/UN-Water 2018 *The United Nations World Water Development Report 2018: Nature-Based Solutions for Water.* UNESCO, Paris.

Zampori, L., Dotelli, G. & Vernelli, V. 2013 Life cycle assessment of hemp cultivation and use of hemp-based thermal insulator materials in buildings. *Environmental Science and Technology* **47** (13), 7413–7420.

First received 27 June 2019; accepted in revised form 11 November 2019. Available online 3 December 2019

doi: 10.2166/bgs.2020.932

A review of nature-based solutions for urban water management in European circular cities: a critical assessment based on case studies and literature

Hasan Volkan Oral[a], Pedro Carvalho[b], Magdalena Gajewska[c], Nadia Ursino[d], Fabio Masi[e], Eric D. van Hullebusch[f], Jan K. Kazak[g], Alfonso Exposito[h], Giulia Cipolletta[i], Theis Raaschou Andersen[j], David Christian Finger[k], Lena Simperler[l], Martin Regelsberger[m], Vit Rous[n], Matej Radinja[o], Gianluigi Buttiglieri[p,q], Pawel Krzeminski[r], Anacleto Rizzo[e], Kaveh Dehghanian[s], Mariyana Nikolova[t] and Martin Zimmermann[u]

[a] Energy Politics and Market Research Center (EPPAM) and Civil Engineering Department, Istanbul Aydin University, Istanbul, Turkey E-mail: volkanoral@aydin.edu.tr

[b] Department of Environmental Science, Aarhus University, Frederiksborgsvej 399, Roskilde, 4000, Denmark

[c] Faculty of Civil and Environmental Engineering, Gdansk University, 11/12 Gabriela Narutowicza Street, 80-233 Gdańsk, Poland

[d] Dept. ICEA, University of Padova, via Loredan 20, 35131 Padova, Italy

[e] IRIDRA Srl, Via Alfonso La Marmora, 51, 50121, Florence, Italy

[f] Université de Paris, Institut de Physique du Globe de Paris, CNRS, F-75005 Paris, France

[g] Department of Spatial Economy, Wrocław University of Environmental and Life Sciences, ul. Grunwaldzka 55, 50-357 Wrocław, Poland

[h] Department of Economic Analysis, University of Seville, Ramón y Cajal 1, 41018, Seville, Spain

[i] Department of Materials, Environmental Sciences and Urban Planning-SIMAU, Polytechnic University of Marche (UNIVPM) – INSTM Research Unit, Via Brecce Bianche 12, 60121 Ancona, Italy

[j] Research Center for Built Environment, Energy, Water and Climate, VIA University College, 8700 Horsens, Denmark

[k] The Sustainability Institute and Forum (SIF), School of Science and Engineering, University Reykjavik, Menntavegur 1, 101 Reykjavik, Iceland

[l] University of Natural Resources and Life Sciences, Institute for Sanitary Engineering and Water Pollution Control, Vienna (BOKU), Muthgasse 18, A-1190 Vienna, Austria

[m] Technisches Büro für Kulturtechnik, Marburger Gasse 11 8200 Gleisdorf, Austria

[n] Faculty of Environmental Sciences, Czech University of Life Sciences, Kamycka 129 165 21 Prague 6 – Suchdol, Czech Republic

[o] Faculty of Civil and Geodetic Engineering, University of Ljubljana, Jamova 2, 1000 Ljubljana, Slovenia

[p] Catalan Institute for Water Research (ICRA), C. Emili Grahit 101, 17003 Girona, Spain

[q] Universitat de Girona, Girona, Spain

[r] Section of Systems Engineering and Technology, Norwegian Institute for Water Research (NIVA), Gaustadalléen 21, N-0349 Oslo, Norway

[s] Civil Engineering Department, Istanbul Aydin University, Istanbul, Turkey

[t] Geodesy and Geography, National Institute of Geophysics, Sofia, Bulgaria

[u] ISOE – Institute for Social-Ecological Research, Frankfurt am Main, Germany

doi: 10.2166/bgs.2020.932

Abstract

Nature-based solutions (NBS) can protect, manage and restore natural or modified ecosystems. They are a multidisciplinary, integrated approach to address societal challenges and some natural hazards effectively and adaptively, simultaneously providing human well-being and biodiversity benefits. NBS applications can be easily noticed in circular cities, establishing an urban system that is regenerative and accessible. This paper aims to offer a review on NBS for urban water management from the literature and some relevant projects running within the COST Action 'Implementing nature-based solutions for creating a resourceful circular city'. The method used in the study is based on a detailed tracking of specific keywords in the literature using Google Scholar, ResearchGate, Academia.edu, ScienceDirect and Scopus. Based on this review, three main applications were identified: (i) flood and drought protection; (ii) the water-food-energy nexus; and (iii) water purification. The paper shows that NBS provide additional benefits, such as improving water quality, increasing biodiversity, obtaining social co-benefits, improving urban microclimate, and the reduction of energy consumption by improving indoor climate. The paper concludes that a systemic change to NBS should be given a higher priority and be preferred over conventional water infrastructure.

Key words: climate change resilience, nature-based solutions, stormwater, urban water, wastewater treatment

INTRODUCTION

According to the 2018 Intergovernmental Panel on Climate Change (IPCC) report, global climate change will cause irreversible harm to humans, the built environment and the biosphere (IPCC 2015). In particular, the depletion and degradation of pristine water resources is expected to affect significantly human and environmental health. In addition, the rapid increase of urban areas, resulting in a higher demand for water resources as well as disruption of the natural water cycle, accentuates the importance of sustainable and resilience-based water management. Hence, it is essential for urban water management (UWM) to be an integral part of urban planning. Moreover, land use decisions affect water supply and wastewater system designs and operation, as well as measures needed for managing stormwater runoff. Furthermore, an urban infrastructure system requires energy, which in turn, typically requires water (Loucks & Van Beek 2017). Consequently, water is one of the key elements of the United Nations Sustainable Development Goals (SDGs), alone or interlinked with different aspects. For instance, several of the 17 objectives are strongly connected to urban farming and call for an economical utilisation of assets, environment rebuilding, biodiversity, carbon sequestration, feasible catchment management and soil management (Keesstra *et al.* 2016).

Urban water refers to all water that is present in urban environments which includes natural surface water, groundwater, drinking water, sewage, stormwater, flood overflow water and recycled water (a third pipe, stormwater harvesting, sewer mining, managed aquifer recharge, etc.). Furthermore, a wide range of techniques can solve urban water-related problems, for example, improving water use efficiency and water demand reduction techniques, water sensitive urban design (WSUD) techniques, living streams, environmental water and protection of natural wetlands, waterways and estuaries in urban landscapes (water.gov.au, 2017). Larsen & Gujer (1997) defined UWM as a combination of water supply, urban drainage, wastewater treatment and water-related sludge handling. Accordingly, UWM includes the plan, design and operation of infrastructure to secure drinking water and sanitation, the control of infiltration and stormwater runoff, recreational parks and the maintenance of urban ecosystems.

Sustainable urban development includes a holistic management approach consisting of the water-energy-food nexus, land use and the diversification of water sources for reliable supplies (Kalantari *et al.* 2018). Further, integrated urban water management (IUWM) provides a framework and

doi: 10.2166/bgs.2020.932

objective for planning, designing, and managing urban water systems. Moreover, IUWM is a flexible process that responds to change and enables stakeholders to participate in, and predict the impacts of development decisions. Consequently, adequate IUWM includes the environmental, economic, social, technical and political aspects of UWM. It enables better land use planning and the management of its impacts on freshwater supplies, treatment and distribution; wastewater collection, treatment, reuse and disposal; stormwater collection, use and disposal; and solid waste collection, recycling and disposal systems. Accordingly, it makes urban development part of integrated basin management, which is oriented toward a more economically, socially and environmentally sustainable mixed urban-rural landscape (Loucks & Van Beek 2017; Kalantari *et al.* 2018; Arabameri *et al.* 2019). IUWM also aims to help cities progress towards a circular economy, thus closing the loop of water resource circulation, and helps to limit the discharge of liquid waste and the constantly growing need for additional water resources (High level Panel on Water 2018).

As a result of increasing urban areas, the interaction of many factors such as demographic, economic, political, environmental, cultural and social factors creates challenges related to the use and management of water resources. Several of these issues can be addressed with nature-based solutions (NBS). NBS aim to protect, sustainably manage, and restore natural or modified ecosystems. NBS address societal challenges effectively and adaptively, simultaneously providing human well-being and biodiversity benefits (IUCN 2018). NBS also have the potential to underpin a sustainable water management strategy (FAO 2018).

This study is conducted under the European Cooperation in Science and Technology (COST), which funds the research Action 'CA17133 – Implementing NBS for creating a resourceful circular city'. The circular city model recognises the importance of organising the city's systems in an analogy to the organisation of natural systems and it incorporates the principles of the circular economy, establishing an urban system that is regenerative and accessible (Girard & Nocca 2019). This study aims to offer a brief review on NBS for UWM, together with a description of some relevant projects running within the action. In this COST Action, the definition of a common language and understanding across disciplines is seen as a crucial success factor, while circular economy (CE) concepts are seen as a key approach and NBS or green infrastructure (GI) solutions are seen as core elements of the toolbox (Langergraber *et al.* 2019). Our working group has focused on the implementation of a safe and functional water cycle within the urban biosphere, where wastewater needs to be streamlined as a source of nutrients, hazardous pollutants that should be controlled (e.g., heavy metals or emerging organic contaminants), heavy metals being phytomined, the treated water looped back for irrigation, and recreational purposes should be considered side by side with sanitation, water supply or stormwater management. Furthermore, we critically appraise the established centralised water flow, defining available resources within the water flow and risk assessment on urban water, NBS for stormwater management and wastewater treatment.

The main research question addressed in this paper is 'How can NBS be integrated with the sustainable UWM?' To answer this question we followed two parallel approaches: (i) a traditional literature review targeting a set of different subtopics, coupled with (ii) an overview of case studies from projects running within the framework of the COST Action. By combining both, we wish to provide not only the most complete overview of the current existing knowledge but also to discuss and challenge the current existing frameworks for NBS implementation. Therefore, the aim of this paper is to define the challenges, present benefits and future trends, provide an overview of the usage of NBS for UWM and to offer implementation recommendations for urban water utilisation towards circular cities.

The paper is organised as follows. The 'methodology' section presents both the framework for the literature review and the selection criteria of relevant case studies. The next section describes the review through existing NBS tools for sustainable water management, subdivided into the sections 'stormwater management', 'flood protection and risk management', 'implementation of blue-green infrastructures', 'urban water in the field of food, water, and energy ecosystem', and 'urban water pollution control: constructed wetlands'. Following on, a section describes some case studies, linking them with the existing literature. The final section offers a brief discussion, and some concluding remarks are provided to point the way forward for an increased implementation of NBS for UWM.

METHODOLOGY

Owing to the broad scope of the topic, different levels of implementation of NBS and availability of international peer-reviewed literature for certain subtopics, we propose a combined approach where both existing literature and case studies were reviewed using different criteria. This section is divided into two subsections. In the first subsection, we present the details of the literature survey to collect data of relevant international peer-reviewed journals, while in the second we describe the criteria for selecting relevant case studies important for the current review.

Literature review approach

The literature survey was performed independently by different sub-groups of authors involved in this work. Therefore, the details of the literature search are described in the next paragraphs per each subsection.

For the stormwater management section, the literature was searched in Google Scholar, ResearchGate, Academia.edu, ScienceDirect and Scopus by using the key words 'stormwater management' AND 'nature based solutions', 'stormwater management' AND 'historical development', 'climate change' AND 'resilience'. A total of 40 manuscripts (i.e., 10 Google Scholar, 2 ResearchGate, 2 Academia.edu, 15 ScienceDirect, 4 Scopus and 7 other publications were retrieved by cross-checking references from the initial retrievals from the databases) were retrieved and revised. After the first screening of the abstracts, 19 papers were disregarded and 11 were read and are presently discussed in this paper.

For the flood protection and risk management section, the literature was searched in Google Scholar, ResearchGate, ScienceDirect and Scopus by using the key words 'flood' 'risk management' AND 'nature based solutions'. A total of 34 manuscripts (i.e., 12 Google Scholar, 3 ResearchGate, 10 Science Direct, 3 Scopus and 6 other publications were retrieved by cross-checking references from the initial retrievals from the databases) were retrieved and revised. After the first screening of the abstracts, 23 papers were disregarded and 11 were read and are presently discussed in this paper.

For implementation of blue-green infrastructures section for flood protection and risk management section, the literature was searched in Google Scholar, ResearchGate, ScienceDirect and Scopus by using the key words 'flood', 'risk management', AND 'nature based solutions'. A total of 45 manuscripts (i.e., 10 Google Scholar, 12 ResearchGate, 8 ScienceDirect, 10 Scopus and 5 other publications were retrieved by cross-checking references from the initial retrievals from the databases) were retrieved and revised. After the first screening of the abstracts, 26 papers were disregarded and 19 were read and are presently discussed in this paper.

For the urban water pollution control section, a total of 60 manuscripts (i.e., 20 Google Scholar, 10 ResearchGate, 15 ScienceDirect, 10 Scopus and 5 other publications) were retrieved by Google Scholar, ResearchGate, ScienceDirect and Scopus by using the key words 'nature based solutions' AND 'urban water pollution control'. Among them, 10 papers were disregarded and 50 were read and discussed.

For the water-energy-food nexus section, the literature was searched in Google Scholar, ResearchGate, ScienceDirect and Scopus by using the key words 'nature based solutions' 'water' AND 'energy' AND 'food nexus'. A total of 36 manuscripts (i.e., 4 Google Scholar, 1 ResearchGate, 12 Science Direct, 2 Scopus and 17 other publications were retrieved by cross-checking references from the

doi: 10.2166/bgs.2020.932

initial retrievals from the databases) were retrieved and revised. After the first screening of the abstracts, 13 papers were disregarded and 23 were read and are presently discussed in this paper. This section also generated two supplementary tables: Table S1 overviews the existing NBS and their link with the water-food-energy nexus and Table S2 compares groundwater-based natural infrastructure solutions with grey infrastructure.

Case studies selection criteria

International projects in which the CA1733 action members are directly involved dealing with NBS and sustainable water management were selected as case studies for this article. The data related to these case studies were obtained from the researchers involved in both the projects and COST Action. Figure 1 shows the geographical location of the case studies, further detailed in the section 'Case studies'.

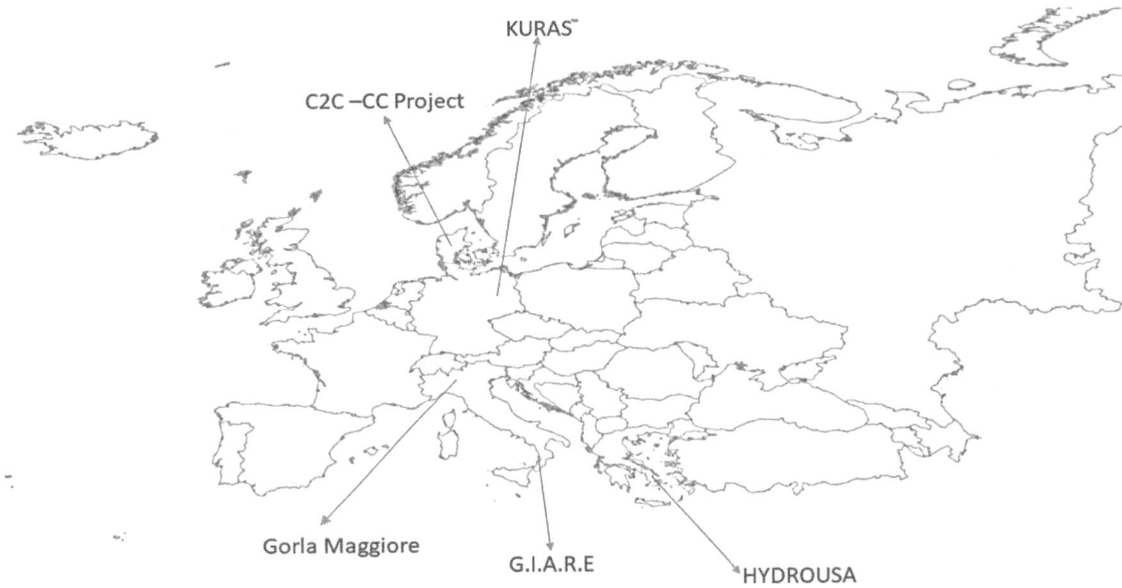

Figure 1 | Location of reviewed case studies within the COST Action.

REVIEW OF THE EXISTING NBS TOOLS FOR SUSTAINABLE WATER MANAGEMENT

In this time of anthropogenic climate change, urban regions around the world face natural disasters such as heat islands, droughts and floods, as well as urban pressures, for instance, air and water pollution along with resource management inefficiency. Consequently, the sustainable development of urban areas has resulted in decision-makers being caught in challenging situations, while simultaneously having to solve the problem of the excess of one resource and the lack of others. Therefore, based on individual cases, it seems rational to consider the possibility of implementing the concept of the circular economy in addition to connecting two problems – instead of defining them and seeing one aspect as the solution for other elements of a healthy urban socio-environmental system.

In this section, we group the sustainable water management under five categories as: (i) stormwater management, (ii) flood protection and risk management, (iii) implementation of blue-green infrastructures, (iv) urban water in the field of food, water and energy ecosystem and (v) urban water pollution control: constructed wetlands.

Stormwater management

In recent years, stormwater management has become an increasingly multidimensional and multidisciplinary issue. Moreover, stormwater presents very distinct qualitative and quantitative characteristics from domestic sewage. It is recognised as the most important source of heavy metals, whereas wastewater constitutes the main source of organic and nitrogenous pollution (Bavor *et al.* 2001; Eriksson *et al.* 2007; Barbosa *et al.* 2012).

In many countries, separate sewer network systems are predominant, and most rainwater networks discharge rainwater directly to receiving waters, without any purification, which is a serious threat to the quality of such water. This is particularly dangerous for small watercourses flowing through cities for which rapid discharge from rainwater drainage systems exceeds the hydraulic capacities, and the introduced pollution load is a serious threat. Further, until the 1990s, it was believed that the best solution to the rainwater problem in cities should be drainage, i.e., efficiently collecting and discharging stormwater to receiving waters (Figure 2).

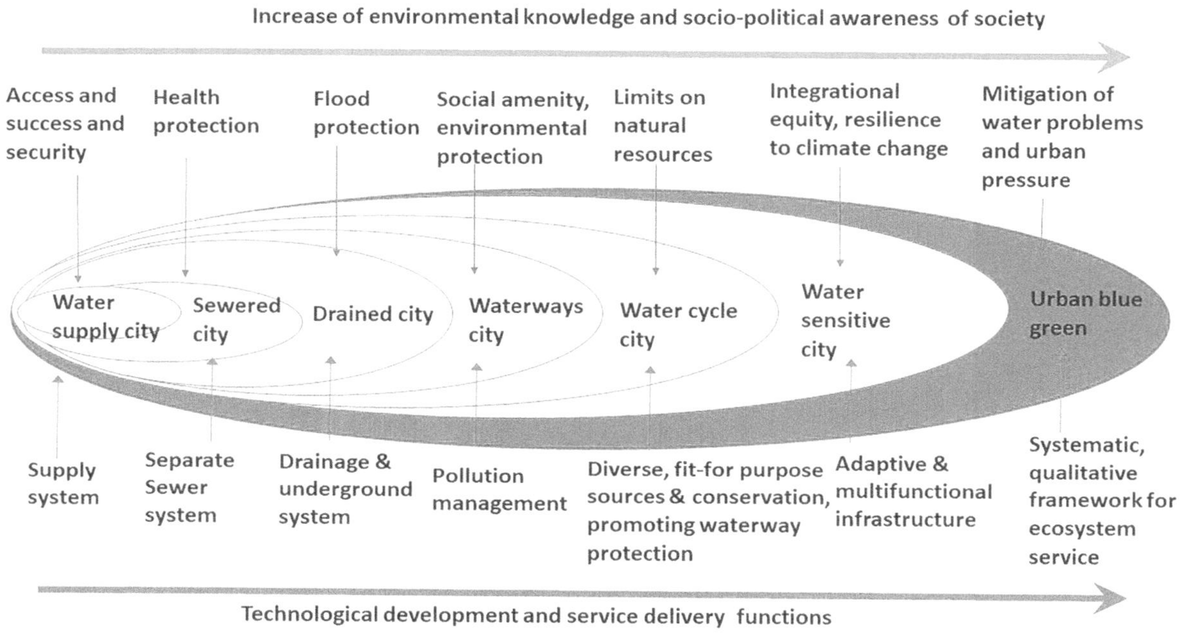

Figure 2 | Historical development of water supply and management (modified from Brown *et al.* (2009) and Blue Green Solution (2017)).

However, progressing urbanisation is inevitably connected to replacing the natural land cover with impermeable surfaces, which leads to increased surface runoff. Additionally, climate change is leading to more volatile rainfall patterns with an increasing number of extreme events, thereby causing frequent overloading of the drainage systems. As a result, floods are occurring, especially in central city districts with a high level of impervious surfaces. Such events, referred to as pluvial flash floods, are followed by long dry spells. For example, over the last 18 years in Gdańsk, Poland, more than four rainfall events with a 100-year return period (i.e., over 100 mm/day) have occurred. On 14/15 July 2016, 160 mm of rain fell within 14 hours, exceeding the total rainfall of two months. On the other hand, as mentioned above, long periods without precipitation are also causing functional problems for cities. Thus, the lack of stored rainwater increases the need for watering urban green areas with irrigation systems. Such approaches require both natural resources and financial support, thereby leading to their unsustainability (Wojciechowska *et al.* 2015).

doi: 10.2166/bgs.2020.932

Despite the risks that water can pose in urban spaces, it is an integral part of the city and a vital resource for the residents. From the human health perspective, it is necessary to integrate water in the urban layout. Therefore, a modern approach to the urban planning of the so-called WSUD assumes the use of the most natural technological solutions, the so-called eco-engineering. We count green roofs, bioretention systems, 'rainforests' and hydrophyte systems that combine the function of purification and retention and provide many ecosystem services (ES), including biodiversity and returning rainwater to the local water cycle by evapotranspiration. The natural ground cover would only have 10% runoff with 40% via evapotranspiration and 50% through infiltration while the impervious cover would have 55% runoff with 30% evapotranspiration and 15% infiltration (US EPA 2003).

As presented above, existing water management systems are not sufficient in many cases, and the need to solve the problem of quantity and quality of water exists in order to implement the concept of an urban circular economy. The synergy of constantly growing urban areas with impervious surfaces and pollution associated with human activities, and climate change with an increasing number of meteorological extremes, requires a new approach for cities to become more resilient to socio-environmental pressures (Figure 3).

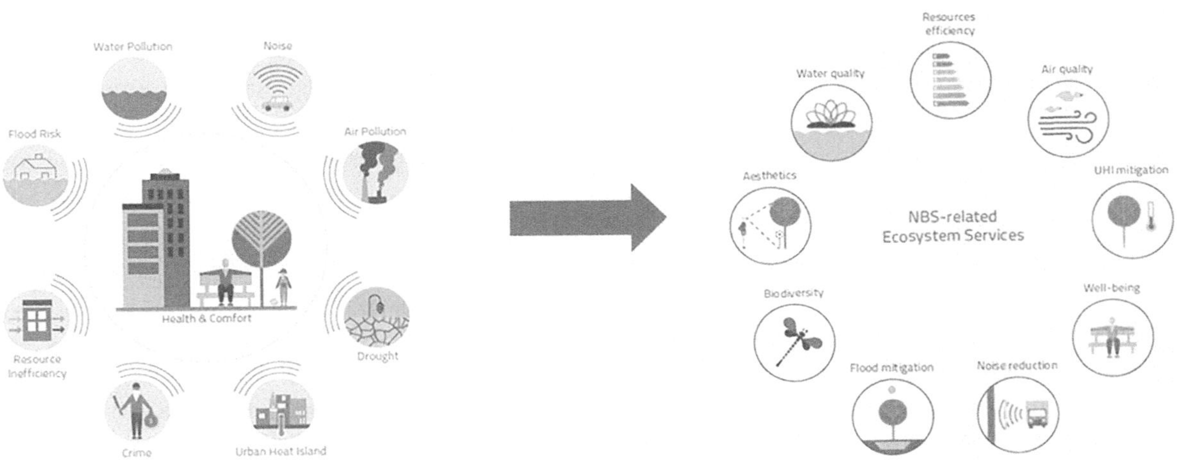

Figure 3 | Identified water problems and urban pressures and mitigation options by the application of NBS (Blue Green Solution 2017).

Therefore, based on the identified challenges, there is an urgent need to support the implementation of NBS in cities in order to contribute to climate change adaptation by reducing their vulnerability to environmental threats. NBS allow mimicking pre-development hydrologic regime and detaining runoff close to its source following the principle of low-impact development (Coffmann 1999; Bavor *et al.* 2001; Wong & Brown 2009; Hoyer *et al.* 2011) and use plants to later return the water to the local water cycle through evapotranspiration, thus supporting the plants in dry periods. Therefore, NBS become an essential feature of urban resilience managing stormwater, contributing to urban cooling through evapotranspiration and alleviating urban heat island effects while supporting urban green with local water resources.

Flood protection and risk management

Ecosystems, depending on their management, can either contribute to the problem or provide effective NBS for flood risk reduction or climate change mitigation and adaptation (Cohen-Shacham *et al.*

2016). At the same time, the implementation of NBS depends on the state and capacity of ecosystems to provide particular regulating services (flood, erosion, climate). Their spatial dimensions provide a basis for land use management and urban planning decisions in accordance with an ecosystem-based approach for flood risk management and other aspects of urban environmental management (Szopińska *et al.* 2019). On the other hand, there are other, potentially very cost-effective ways of achieving flood protection by tapping into nature's own capacity to absorb excess waters. Consequently, NBS implementation aims at preventing natural disasters to make urban areas safe and resilient, which can be achieved in combination with technological and engineering solutions if necessary.

Planning infrastructures to manage flood risk is related to connectivity (Parson *et al.* 2015), circularity (Kirchherr *et al.* 2017; Comino *et al.* 2018; Keesstra *et al.* 2018) and finding a balance between natural and urban elements (Gaines 2016). Moreover, in a fast developing city, the loss of circularity is often associated with the altered hydrological cycle, implying that water is not a natural, valuable resource, but rather a threat to the urban environment, when it flows at rates different from those of natural paths, from/toward locations that are functional to the development of human activity rather than to the environmental dynamics, through man-managed (often fast) connections, with quality standards far from those provided by natural water bodies (EPA 2005).

Consequently, the loss of circularity in the altered natural water cycle is derived from the reduction of soil infiltration capacity and resulting in fast surface runoff. The fact that the natural water cycle is replaced by the urban water cycle threatens soil, channelised urban drainage systems, receiving water bodies and downstream cities. Furthermore, the wash-off of pollutants from anthropogenic catchments poses a threat to the receiving water bodies and their biomes. The loss of infiltration and uncontrolled leakage from sewage threaten groundwater and connected surface water bodies. Subsequently, the resources, politics and awareness affect the socio-environmental dynamics and determine whether the socio-hydrological system will undergo irreversible decline or be self-sustainable (Ursino 2019).

NBS, in this context, is meant to partially recover the pre-development water fluxes and water quality, thus reducing the flood risk (WWAP 2018). Therefore, the use of NBS in this context is strongly related to the well-known concept of sustainable urban drainage, known in the literature with different key words, such as sustainable drainage systems (SuDS), WSUD or low impact development (LID), as reviewed by Fletcher *et al.* (2015). All these concepts aim to restore the water cycle within an urban catchment, from post-development back to the pre-development state (Fletcher *et al.* 2013). Thus, based on site-specific characteristics and the aim of implementation (to recover original functionality of the urban catchment or address specific issues linked to water management and risk control), NBS alone may not be able to re-establish complete circularity of the natural water cycle but rather provide multiple services to the community (e.g., mitigate flood and drought risk, affect local climate conditions, increase amenity and biodiversity). Further, based on the scale at which NBS are integrated into the so-called GI, different benefits can arise (Golden & Hoghooghi 2018). For instance, Zhang *et al.* (2019) investigated how NBS across facility, catchment and continental scales differently impact the hydrological, water quality and bioecological benefits.

Implementation of blue-green infrastructures

One of the most common ways to implement NBS is by the so-called blue-GI. Blue-green infrastructures are key elements in the holistic planning of (future) urban regions (Winker *et al.* 2019). Accordingly, blue-green infrastructures create strategically planned networks of (artificial) natural spaces in cities (Bundesamt für Naturschutz 2017). Therefore, the use of NBS seeks to minimise the effects of climate change on urban areas and create various ecosystem services with benefits

doi: 10.2166/bgs.2020.932

for the society, environment and economy. NBS can help create natural circumstances in urban areas for 'alleviating urban pressures and achieve resilience to climate change' (Maksimovic *et al.* 2017).

Blue-green infrastructures establish multifunctional structures as diverse green spaces in combination with elements of WSUD (Winker *et al.* 2019) to strengthen urban sustainable development. Accordingly, 'green' infrastructural elements take essential roles in creating a healthy microclimate in cities. For instance, trees reduce flood risks and the effect of urban heat islands and expand shading while pocket parks (and streams) aesthetically attract citizens and provide space relieving mental aspects of urban pressure (Maksimovic *et al.* 2017). 'Greening' transforms cities by unsealing surfaces and is applied on building structures to lower buildings' energy level by natural cooling, which saves costs and works in aesthetical ways. Further, green rooftops have a multifunctional use within blue-green infrastructures from urban gardening to collecting spaces for rainwater. As precipitation is a scarce resource and floods and droughts will accumulate due to climate change, cities can adapt WSUD strategies which focus on managing all water streams within the city. In addition, a water supply from rainwater, stormwater and treated wastewater from a sustainable blue infrastructure for cities can relieve or replace grey infrastructure (Depietri & McPhearson 2017). For example, natural or close to natural ways of flood risk prevention such as sponge cities are more sustainable than flood walls. In combination, blue-green infrastructures provide health benefits for society and relieve the pressure on the environment and urban space. Furthermore, blue-green infrastructures are more cost-effective than the current predominant urban infrastructures. As such, a long-term sustainability approach can be pursued with regard to the design of cities of the future.

The urban blue-green infrastructures provide various valuable regulating ecosystem services in respect to global climate regulation by reduction of greenhouse gas concentrations through carbon storage and sequestration (Kazak *et al.* 2016), water flow maintenance and flood protection (Szewrański *et al.* 2018), micro and regional climate regulation (Kołecka *et al.* 2018; Ziemiańska & Kalbarczyk 2018) and improvement of air and water quality (Lakatos *et al.* 2012; Dąbrowska *et al.* 2017; Bawiec 2019). Consequently, creating well-designed built environments rich in ecosystem services provides various options for mitigation and adaptation of urban areas for the impact of climate change. Most of the adaptation measures in cities depend mainly on particular urban planning solutions and public regulations. Therefore, based on the technological solutions, local authorities can improve urban development processes by decision support systems, which effectively suggest suitable solutions in the case of many domains of environmental management (Kazak & van Hoof 2018). Identification and consideration of the dependency of the local population on the particular ecosystem services in the living areas make the valuation of the ecosystem services an important factor in sustainable landscape planning and territorial integration policymaking (Borisova 2013; Świąder *et al.* 2018).

As mentioned above, the implementation of blue-green infrastructures does not only solve the problem of water management in cities, but it supplies much more influential ecosystem services on increasing urban resilience to socio-environmental challenges. These ecosystem services can be assessed and mapped for better understanding of the environmental carrying capacity in the land management system to cope with flood hazard at all levels – region, basin and settlement (Boyanova *et al.* 2014; Świąder 2018). In some cases, the ecological boundaries, in terms of the area providing ecosystem services to the cities, exceed their administrative boundaries up to 1,000 times (Folke *et al.* 1997). At the same time, cities rely heavily on the capacity of the ecosystems in the urban environment provided by the green and blue areas. Thus, the interaction between biophysical and geophysical processes determines the potential capacity of natural capital to provide regulating ecosystem services. The water flow can be influenced by several natural processes and functions of the ecosystems, which contribute to the absorption of water and therefore reduce surface runoff or vice versa. The main factors of the capacity for water retention are the vegetation cover, the soil structure and texture, the presence of bare land or water bodies, the slopes and the land cover in the territory. In the study by Nikolova & Nedkov (2018), the flood regulation supply capacity was assessed by an Index of Capacity for Water Retention of urban ecosystems (as defined by Zhiyanski *et al.* 2017). The assessment of flood regulation services is carried out in four main steps according to the methodological framework for ecosystem services assessment developed by Burkhard *et al.* (2012):

1. identification of the urban ecosystems with potential to provide flood regulation;
2. selection of indicators for ecosystem services assessment;
3. quantification of the ecosystem services indicators;
4. assessment and mapping of flood regulating urban ecosystem services.

The results of such assessment show that the water retention capacity of residential, industrial and public areas is lower, while urban green areas have higher potential. Thus, detailed assessment gives decision-makers the exact information about the impact of future actions on biocapacity and the ecological footprint of human activity.

Urban water in the field of food, water and energy ecosystem

Liquete *et al.* (2016) as well as Leigh & Lee (2019) indicated that the future of urban water systems is shifting towards resource oriented, integrated, sustainable, distributed and NBS. Accordingly, wastewater treatment will be replaced by the production of goods. Further, one optimised system will allow multiple targets to be reached, instead of having a separate infrastructure for every purpose. This should give treated wastewater access to everybody. Multiple targets besides water treatment can be the production of fertilisers, provision of urban green, enhancing biodiversity and cooling, to name just one possible set. These targets must be defined during the concept phase in a case by case approach and their fulfilment must be measurable. According to present knowledge, this is the way forward to eliminate present untreated wastewater releases, a target set in SDG 6: clean water and sanitation (UN 2015).

NBS can help face these challenges by providing the means for cities to successfully achieve long-term sustainability in the use of resources (e.g., energy, water, land) and increase urban resilience to climate change (Maes & Jacobs 2017). Nevertheless, water-energy-food nexus relationships are complex and poorly understood, especially in urban environments, thus leading to significant potential risks. However, there are benefits if society is able to manage them adequately (Bennett *et al.* 2016). Further, Bennett *et al.* (2016) address the water-energy-food nexus and natural infrastructure investment on the entire watershed scale, taking large-scale infrastructure investment programmes into account, thus going beyond the city boundaries. Consequently, the implementation of NBS in urban areas can benefit from the water-energy-food nexus at local scales to efficiently manage natural resources for the optimal ecosystem services' delivery. Nevertheless, to the best of our knowledge, there are no literature reviews focused on the water-energy-food nexus in urban areas and how multi-functional NBS may help manage this nexus to improve the usage efficiency of these resources, thus helping to achieve long-term sustainability of cities. Some recent studies, such as those of Hansen *et al.* (2015), Lafortezza *et al.* (2018), Krauze & Wagner (2019) and Keesstra *et al.* (2018), describe NBS with multifunctional targets and affecting the water-energy-food nexus in urban areas. At the European scale, besides the main reports from the EC (2013, 2015), recent studies have analysed NBS applications in urban environments: Faivre *et al.* (2017) focus on NBS to address social, economic and environmental challenges in EU areas; Kabisch *et al.* (2016) review NBS for climate change adaptation in urban areas; Nikolaidis *et al.* (2017) study new approaches to improve regulatory instruments and demonstrate the long-term value of NBS; Raymond *et al.* (2017a) develop a framework for assessing and implementing the co-benefits of NBS in urban areas; Russo *et al.* (2017) review NBS based on edible GI for better management of the water-energy-food nexus; and the reports from

doi: 10.2166/bgs.2020.932

the Naturvation project (Bockarjova & Botzen 2017; da Rocha *et al.* 2017; Hanson *et al.* 2017), which review the different dimensions of NBS implemented in urban areas, including those related to a more efficient use of natural resources and the nexus between water, energy and food in NBS.

Most frequently, NBS are designed for: (1) urban water regeneration; (2) watershed management; (3) ecosystem restoration; (4) increasing the sustainable use of matter; (5) generation of renewable energy; and (6) increasing carbon sequestration. Likewise, European authorities (EC 2013, 2015) have highlighted the multifunctional benefits of NBS to improve resource efficiency in urban areas. Among these solutions, we find: (1) urban agriculture for local food production; (2) water regeneration; (3) green roofs for climate adaptation; (4) higher energy and water efficient use; (5) regeneration of abandoned land by afforestation; (6) food production; (7) rain gardens for stormwater regulation; and (8) the use of permeable surfaces and vegetation for run-off control. Tables S1 and S2, given as the Supplementary material, present the examples of relevant NBS related to the water-energy-nexus. Finally, one of the main challenges in the topic is the assessment of the performance and impacts of NBS in addressing the objectives of higher resource efficiency and resilience in urban areas. The assessment schemes have been developed to measure performance and impacts through different indicators: Mapping and Assessment of Ecosystems and their Services (MAES) (Maes *et al.* 2013), Knowledge and Learning Mechanism on Biodiversity and Ecosystem Services (EKLIPSE) (Raymond *et al.* 2017b) and the Smart City Performance Measurement Framework (CITYkeys) (Bosch *et al.* 2017). In addition to the examples of relevant NBS related to the food-energy-nexus, the application of groundwater-based natural infrastructure solutions and comparison with the grey infrastructure also exist. Table S2 explains the function, goal and solution, which are the outcomes of the comparison.

Urban water pollution control: constructed wetlands

Urban water pollution control nowadays is predominantly carried out as an 'end of the pipe' solution with highly intensified wastewater treatment systems in order to protect downstream freshwaters from contamination and eutrophication (Finger *et al.* 2013). Yet, in addition to the benefits related to management of stormwater, flood protection and efficient use of resources in a water-energy-food nexus discussed in the previous sections, NBS offers an untapped potential for urban water pollution control. The treatment potential of NBS depends, among other factors, on the type of NBS used (infiltration basin, constructed wetland, raingarden, etc.), quantity and quality of water to be treated, and local conditions (climate, precipitation patterns, etc.).

In the concepts of GI, LID and sustainable drainage systems, water pollution controls are provided by the so-called planted/unplanted biofiltration systems. According to the definition of Fonder & Headley (2013), planted (surface) systems are a type of constructed wetlands (CWs). Among the various types of NBS, CWs are the most common and accepted NBS for pollution control nowadays, and they can be used in cities, especially for Masi *et al.* (2018):

- rainwater treatment;
- combined sewer overflow treatment;
- polishing of the outflow from existing wastewater treatment plants, including for the treatment of contaminants of emerging concern (CEC);
- greywater treatment.

In respect to water quantity and quality, stormwater presents different qualitative and quantitative characteristics compared to domestic sewage. It is recognised as the most important source of heavy metals, whereas wastewater constitutes the main source of organic and nitrogenous pollution (Barbosa *et al.* 2012). On the other hand, the quality of stormwater can vary greatly in time and between locations, especially in urban areas where over 650 substances were identified in stormwater

(Eriksson *et al.* 2007). Table 1 displays the classification of five main groups of pollutants that can be encountered in stormwater.

Table 1 | List of main stormwater pollutant types (adapted from Eriksson *et al.* 2007)

Pollutant types	Indicator parameters
Basic parameters	Organic matter (BOD_5, COD), suspended solids, nitrogen, phosphorus, pH
Heavy metals	Zinc, cadmium, chromium (VI), nickel, lead, platinum
Polycyclic aromatic hydrocarbons	Benzopyrene, naphthalene, pyrene
Herbicides	Terbuthylazine, pendimethalin, phenmedipham, glyphosate
Organic compounds	Nonylphenol ethoxylates and degradation products, e.g., nonyl phenol, pentachlorophenol, di-2-ethylhexyl phthalate, 2,4,4′-trichlorobiphenyl (polychlorinated biphenyl 28), methyl-tert-butyl ether
Bacterial indicators	Faecal coliforms (*E. coli*), pathogens (*Pseudomonas aeruginosa*)

Typically, NBS are employed to reduce the levels of traditional pollutants such as total suspended solids (TSS), organic matter, nutrients and also heavy metals. TSS belong to the group of basic pollutants but, at the same time, are classified as being the most dangerous due to their impact, both on the aquatic environment and humans (Makepeace *et al.* 1995; Paschke 2003; Eriksson *et al.* 2007; Madrid & Zayas 2007; Ingvertsen *et al.* 2011; Gasperi *et al.* 2012; Zgheib *et al.* 2012). The concentration of TSS could vary significantly depending on the place of origin (e.g., for streets: TSS ranges from 61 to 320 mg/L; for parking: TSS ranges from 42 to 240 mg/L; and for motorways: TSS is around 200 mg/L (Boogaard 2015). It must also be considered that very often TSS are constituted or covered by organic matter which works as a binding material for the sorption of the above-mentioned emergent pollutants, allowing, therefore, their transport even for a long distance. Therefore, retention of suspended solids has been a primary function of many of the NBS. Typically, CWs can remove up to 88% of TSS, 92% of BOD_5, 83% of COD even after 20+ years of operation (Vymazal 2019). For the nutrients, the removals vary greatly between the systems and are in the range of 46–90% for total phosphorus and 16–84% for total nitrogen (Malaviya & Singh 2012).

In addition to the removal of traditional pollutants such as suspended solids, organic matter and nutrients (Zhang *et al.* 2014a, 2014b; Machado *et al.* 2017; Arden & Ma 2018), CWs are capable of removing organic and inorganic pollutants (Verlicchi & Zambello 2014; Krzeminski *et al.* 2019). Among these, the removal of pesticides (Barceló & Petrovic 2008), heavy metals (Wang *et al.* 2017), pharmaceuticals (Li *et al.* 2014; Zhang *et al.* 2014a, 2014b; Ilyas & van Hullebusch 2019; Zraunig *et al.* 2019) and various other contaminants of emerging concern (CEC) (Imfeld *et al.* 2009; Matamoros *et al.* 2010; Gorito *et al.* 2017; Talib & Randhir 2017) have been explored in the last decade. The observed removal of heavy metals was between 23 and 97% depending on the heavy metal, CWs' type, type of water matrix and others (Malaviya & Singh 2012).

Regarding the CEC, plant-associated NBS have been reported to be crucial for the removal of different CEC (Carvalho *et al.* 2014; Zhang *et al.* 2016; Ilyas & van Hullebusch 2019), which can favour the solution of creating more 'green' in the cities. Therefore, the key removal pathways are the uptake by plants (e.g., carbamazepine), microbial degradation (e.g., ibuprofen, salicylic acid, galaxolide), adsorption and subsequent sedimentation (e.g., triclosan, tetracycline) and photodegradation (e.g., ketoprofen, naproxen, triclosan, diclofenac) (Bi *et al.* 2019).

Although treatment wetlands can achieve high removal of up to 100% of different organic and inorganic chemicals, the removal effectiveness varies significantly and the removal effectiveness of particular compounds may vary depending on the CW design, its operation mode and seasonal

doi: 10.2166/bgs.2020.932

conditions (Verlicchi & Zambello 2014; Zhang *et al.* 2016; Ilyas & van Hullebusch 2019; Krzeminski *et al.* 2019; Zraunig *et al.* 2019). This indicates that for efficient removal, CWs need to be designed and/or adjusted for targeted pollutants. While CWs can be very effective, they are not able to completely remove CEC from the (waste) water. Moreover, hybrid systems combining different types of CWs, or other treatment techniques, might offer increased removal due to the synergistic effects against specific types of pollutants (Garcia-Rodríguez *et al.* 2014; Verlicchi & Zambello 2014; Zhang *et al.* 2014a, 2014b; Ilyas & Masih 2017; Zhang *et al.* 2019). Furthermore, treated water from CWs may be suitable for some reuse applications if they are well designed and maintained (Ilyas & Masih 2017; Arden & Ma 2018; Krzeminski *et al.* 2019). Nevertheless, current knowledge gaps restrict holistic evaluation of CWs' applicability and the estimation of CWs' potential for the removal of CEC.

Regarding the climatic conditions, CWs have been demonstrated to work efficiently in different climatic conditions, but tropical conditions tend to favour treatment performance due to continuous plant growth, extended sunlight exposure and increased microbial activity, these being of particular importance for more recalcitrant pollutants (Zhang *et al.* 2014a, 2014b; Machado *et al.* 2017). However, good comparable removal rates of suspended solids, organic matter and phosphorus are reported for temperate conditions, with only nitrogen removal being affected in a cold climate (Wang *et al.* 2017).

For urban water pollution control, other NBS (and GI elements) can be very effectively used in combination with CWs for purposes such as wastewater source control and separation, water reuse and other means of sustainable sanitation framework (Masi *et al.* 2018). Accordingly, one of the key concepts could be a combination of composting and vermicomposting toilets (Hill & Baldwin 2012; Anand & Apul 2014) and greywater treatment with wetlands or green walls providing the treated water for further reuse. Furthermore, as the space in cities becomes a highly valuable commodity, multipurpose NBS offering other benefits beyond the water treatment and pollution control become a viable alternative (Raymond *et al.* 2017a; Frantzeskaki 2019). Multifunctionality is a key factor, as the water pollution control does not have to be the major role of NBS but can be integrated into stormwater management and biodiversity enhancement.

PROJECTS/CASE STUDIES APPROACH

In spite of the different potential for implementing NBS for UWM, the showcased projects from the COST Action members are only dealing with stormwater management. The applications range from rainwater harvesting in water-scarce areas (e.g., HYDROUSA project in Greece) to the reforestation of watersheds (e.g., Rangárvellir project in Iceland (Keesstra *et al.* 2018)). While both aforementioned cases aim at re-establishing the natural water cycle and increasing natural water retention, the means and purposes differ. Moreover, the Natural Water Retention Measures project, directed by the EU Directorate-General for Environment from 2013 to 2014, aimed to improve the water status on hydromorphology and diffuse pollution, by offering a catalogue of case studies showcasing a broad range of concepts and case studies (nwrm.eu, 2015). However, for effective selection of NBS for stormwater management planning, instruments are still needed. Within the project Concepts for urban rainwater management, drainage and sewage systems (KURAS) in Germany, an integrated planning approach for stormwater management measures was developed considering the other aspects of NBS besides water retention (Matzinger *et al.* 2017). The potential multi-functionality of NBS is an important feature, especially regarding the implementation in circular cities. The Gorla Maggiore water park project in the northern territories of Italy, which includes the use of a water park for NBS applications, and the integrated and sustainable management service for water–energy cycle in urban drainage systems (G.I.A.R.E.) project in southern Italy based on water–energy interaction in Milan, Italy are also summarised in this section. In addition to these two Italian projects, the C2C-CC project, which is carried out in Denmark and includes flood control, water treatment, base-flow and sustainable heat energy applications, is summarised in the section. The main purpose of explaining these five projects in this section is to emphasise the representative of multipurpose NBS implementations for stormwater management.

Project 1: The Gorla Maggiore water park

The Gorla Maggiore water park project, located in Gorla Maggiore, northern Italy, is an urban wetland development focusing on NBS and ecosystem services (The Gorla Maggiore Project 2019). The park aims to protect the city against flooding, improving water quality, increasing biodiversity and obtaining social co-benefits (Rizzo *et al.* 2018). The park, with a total area of approximately 3 ha, comprises sections with different functionalities: (1) stormwater detention for flood prevention (1 ha); (2) domestic water treatment (0.4 ha); and (3) recreational areas (1.3 ha). Furthermore, the combined sewer overflow and excess runoff may be diverted into the park in the case of extreme rainfall events, with an expected reduction of peak flow by 86% and downstream discharge of 8,900 m^3 for events with a ten-year return period. Moreover, it reduces the downstream dissolved organic carbon load by 11.7 t/yr and nitrogen load by 0.4 t/yr, along with social and ecological benefits (Masi *et al.* 2017). In addition, the project demonstrates that the performance and costs of the park are similar or even better than the grey infrastructure for water purification and flood protection (Masi *et al.* 2017).

Project 2: coast to coast climate challenge (C2C-CC project)

The C2C-CC project is a Danish cross-municipality climate adaptation project with 31 partners and 19 supportive partners working to create a climate resilient central region in Denmark (C2-CCC 2018). The sub-project 'Infiltration of surface water through permeable coating' has the primary aim of re-establishing the natural pre-development water cycle and preventing flooding. This is done by harvesting rainwater in the roadbed as the road is made of permeable asphalt. The roadbed is constructed using a gravel mix ensuring a porosity of 30% which can detain the volume of water generated by a 100-year flood. Moreover, the gravel mix removes TSS and heavy metals from the water. Subsequently, the detained water transmits its heat to a geothermal tube, with a length of 800 m, connected to a nearby day-care centre for heating, which is then infiltrated into the soil. Thus, this NBS provides flood control, water treatment, base-flow and sustainable heat energy.

Project 3: HYDROUSA

HYDROUSA aims to revolutionise the water supply chain in Mediterranean regions by demonstrating innovative solutions for water/wastewater treatment and management, which will close the water loops and will also boost their agricultural and energy profile. Relevant to NBS applications, HYDROUSA demonstrates that circular NBS technologies work for wastewater treatment and nutrient recovery, while creating further environmental and societal benefits. The project offers a solution for the problem of rare water reserves in Mediterranean regions in the summer during the high tourism season. The project will not only develop and demonstrate innovative water services, but will also revolutionise the water value chains in Mediterranean areas from water abstraction and use up to sewage treatment and reuse (www.hydrousa.org, 2019). There are five water categories in the HYDROUSA project: rainwater, groundwater, wastewater, water vapour and sea water and the systems defined between these categories are harvesting, recharge and restore, wetlands, vapour condensation and tropical greenhouse. Moreover, biomimicry design concepts and fertigation are being applied to increase the efficiency of the selected NBS. Some of the recovered products of these systems are water for domestic use, irrigation water, biogas, drinking water and salt (Figure 4).

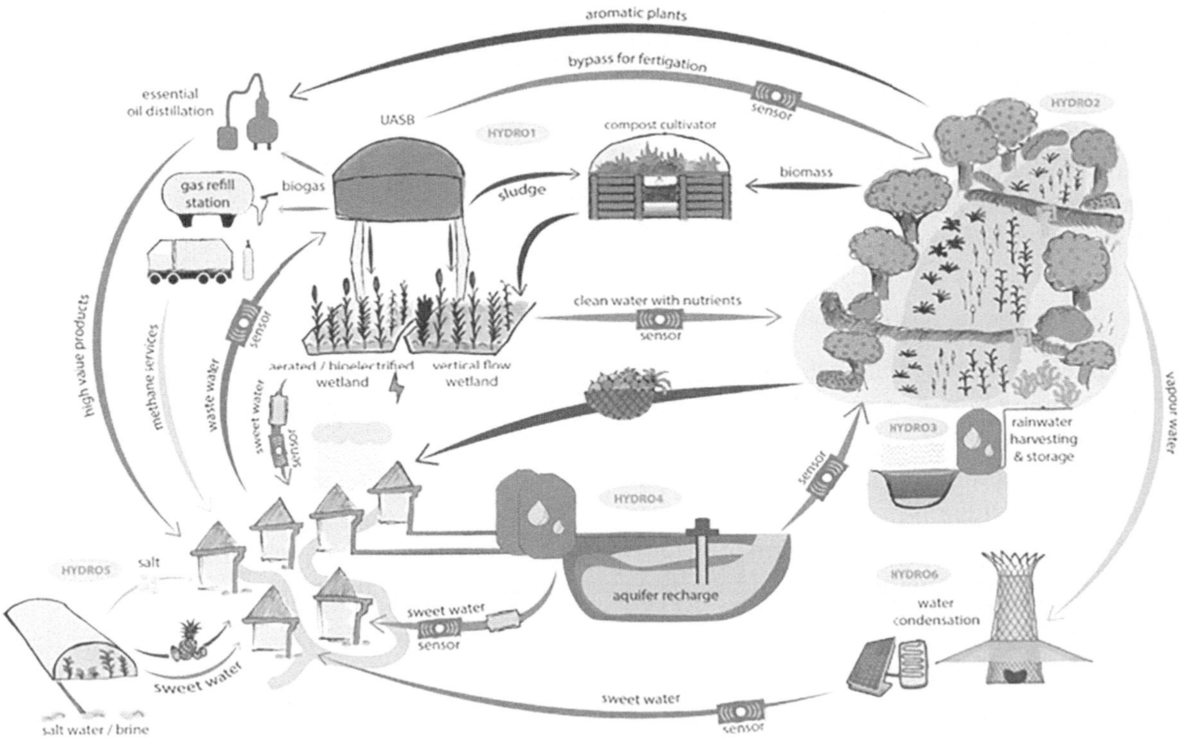

Figure 4 | The HYDROUSA project working principles and the processes (from www.hydrousa.org, 2019).

Project 4: concepts for urban rainwater management, drainage and sewage systems (KURAS)

The aim of this project is to give the answer to the question 'How can the future wastewater discharge, water quality, urban climate and quality of life in the city be improved through intelligently coupled storm water and waste water management?' The project consists of a network of partners from research and industry as well as the city of Berlin decision-makers. KURAS is the elaboration and exemplary demonstration of integrated concepts for a sustainable handling of wastewater and rain-water for urban locations. As mentioned in the Introduction section, NBS aim to protect, sustainably manage and restore natural or modified ecosystems. The KURAS project aims to decrease water consumption after heavy rainfall in the city and enables the sustainable management. Some of the following sub-goals, which are defined, to reach this achievement are as follows (www.watershare.eu, 2019):

- For wastewater disposal companies and operators of municipal sewer networks, which, like Berlin, have a slight gap, options for the adaptation of wastewater infrastructure to climate change and its consequences are being developed.
- Prognosis models are intended to investigate the effects of measures – e.g., to avoid deposits in the sewer system after long periods of dry weather or of mixed water overflows in waters during heavy rainfall – for real Berlin model areas.

Project 5: integrated and sustainable management service for water-energy cycle in urban drainage systems (G.I.A.R.E.)

The main objective of the project, relevant to the aim of NBS, was to develop an integrated approach for sustainable water-energy cycle management in the urban context. In this perspective, a technological platform was implemented in order to both optimise the use of water resources that rely on the urban drainage network such as meteoric waters deriving from the roof of buildings (40% of total urban area) and paved areas, i.e., roads, yards, etc. (35%) and to allow energy saving (Figure 5). For these purposes, experimental activities were conducted on:

- control of inflows to the drainage network;
- control of the polluting load generated;
- thermo-energy benefits;
- potential of rainwater for reuse.

Specific objectives of the project (Figure 5) are listed as follows:

- OR1: Realisation of a compact storm drain prototype device for the treatment of run-off rainwater.
- OR2: Module for management and optimisation of water–energy performance of green roof systems in Mediterranean climate.
- OR3: Urban drainage planning and design service through sustainable technologies to reduce inflows and pollutants.
- OR4: Development of a technological platform for decision-making support for the integrated and sustainable management of the water-energy cycle in the urban drainage system.

An 'Urban Hydraulic Park' was constructed as a demo site at the Vermicelli catchment (University of Calabria) where a green roof with a rainwater harvesting system, permeable pavement, a stormwater filter and a traditional sedimentation tank were connected to a treatment unit. Further, a monitoring and acquisition system was used to analyse the environmental benefits and the hydraulic and thermal efficiency of each unit.

The results of the project showed good hydraulic performance of the green roof concerning stormwater retention in Mediterranean weather conditions (Palermo *et al.* 2019; Piro *et al.* 2018). The hydraulic behaviour of the green roof, permeable pavement and stormwater filter were also analysed by means of a modelist approach (Brunetti *et al.* 2016, 2017; Garofalo *et al.* 2016; Piro *et al.* 2019).

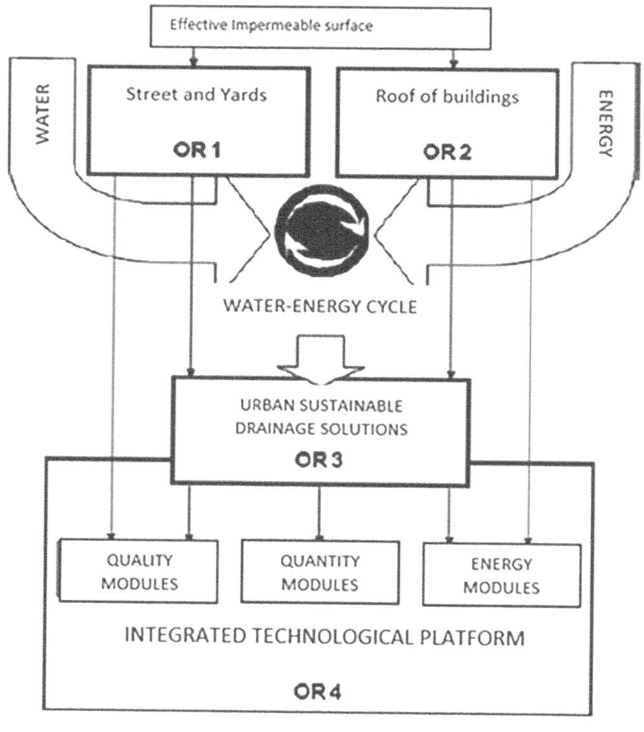

Figure 5 | The working principle of the G.I.A.R.E project (from www.giare.eu, 2019).

doi: 10.2166/bgs.2020.932

Moreover, life cycle assessment (LCA) analysis of the green roof and permeable pavement highlighted the sustainability of these low-impact infrastructures (Maiolo *et al.* 2017).

DISCUSSION AND CONCLUDING REMARKS

Based on the presented literature review and case studies, three main focuses of NBS implementation could be identified: (i) stormwater management, (ii) water-food-energy nexus using water for food and energy production and (iii) water pollution control. The presented overview demonstrates that NBS are not only effective and efficient, but are also largely accepted by people neighbouring such facilities. The present review differentiates from previous studies by combining a literature review with the analysis of case studies involving different NBS applications in European cities. Table 2 presents the review of these case studies that highlight the links between the NBS approach, main services and references. For instance, stormwater management can be implemented by establishing water parks with extended retention basins that withhold rainwater during heavy precipitation events as

Table 2 | Review summary of the case studies

Case studies and implementation years	Description of the NBS	Type of water	Main services	References
The Gorla Maggiore water park, 2013–2015	Constructed wetlands for water pollution control and hydraulic risk management	Combined sewer overflow	Water pollution control; Flood mitigation; Biodiversity increase; Aesthetic/Social benefits	Liquete *et al.* (2016), Masi *et al.* (2017) and Rizzo *et al.* (2018)
HYDROUSA, 2018–present	Constructed wetlands in the food-water-energy nexus; Constructed wetlands for greywater treatment and reuse; Rainwater harvesting systems; Agroforestry	Wastewater; Greywater; Rainwater; Seawater	Reuse of nutrient (N,P) rich treated wastewater; Reuse of treated greywater; Reuse of harvested rainwater; Recreation of high-biodiversity and productive agroforestry site; Removal of emerging organic micropollutants from the water phase and plant uptake in edible plants	The project is still running – www.hydrousa.org (2018–2022)
KURAS, 2016–present	Rainwater harvesting; Decreasing water consumption	Rainwater, Water sewer overflow	Avoiding deposits in the sewer system; Reuse of harvested rainwater	The project is still running – www.kuras-projekt.de
G.I.A.R.E., 2011–2014	Compact storm drain prototype device for run-off rainwater treatment; Green roof systems for support in the management of urban drainage system; Urban drainage planning and design service to inflows and pollutant reduction; Development of a technological platform	Rainwater; Stormwater	Removal of pollutants at storm drain inlet; Management and optimisation of water–energy performance; Hydraulic defence of urban area and control on discharge quality into water bodies; Decision-making support for sustainable management of water–energy cycle in urban drainage system	Brunetti *et al.* (2016), Garofalo *et al.* (2016), Brunetti *et al.* (2017), Maiolo *et al.* (2017), Palermo *et al.* (2019) and Piro *et al.* (2018, 2019)
C2C-CC, 2016–present	Providing flood control; Running water treatment; Waste-flow; Launching sustainable heat energy	Rainwater; Groundwater; Lakes; Rivers; Seawater	Water pollution control; Flood mitigation; Management and optimisation of water–energy performance	Project is still running – https://www.c2ccc.eu/

illustrated in the example of Gorla Maggiore in northern Italy. Such water parks offer protection from floods but also create ecosystems within the cities. Moreover, permeable coating of streets and paths are another way of reducing flood risk in cities. These systems can also produce energy for district heating by simply using the heated surface of paved streets and paths. The Danish project, C2C-CC, is an illustrative example of such a system. The HYDROUSA project investigates options for NBS to manage water resources on Greek islands which experience increased water demand during the tourist season. The KURAS project in Berlin, Germany, focuses on NBS for stormwater and wastewater management in large urbanised areas. Water parks, permeable coating of streets and green roofs function as water retention reservoirs, slowing down the runoff during heavy precipitation events. In some cases, the water stored in these NBS can become available at later points during dry periods, thereby reducing the drought effects. The G.I.A.R.E. in Italy focuses on integrated and sustainable management service for water–energy cycle applications.

Figure 6 illustrates the suggested scheme of sustainable water management in an urban settlement with the case studies HYDROUSA and GORLA MAGGIORE and Figure 7 shows the case studies C2C-CC, KURAS and G.I.A.R.E.

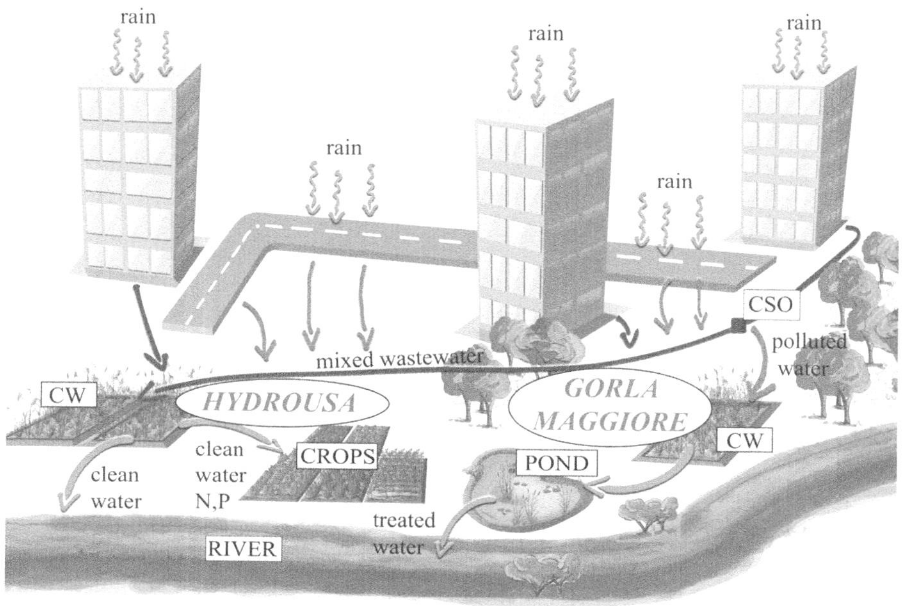

Figure 6 | Transition scheme towards a decentralised and integrated, sustainable water management in an urban settlement with the contributions of the case studies HYDROUSA and GORLA MAGGIORE highlighted (SUDS: sustainable drainage systems, CW: constructed wetlands) (adapted from Masi *et al.* 2018).

The presented review and selected case studies relevant to NBS demonstrate the advantages of NBS both in social and economic terms, i.e., creation of new jobs and saving of energy and resources. Closed-loop recycling of greywater can decrease the amount of potable water used and wastewater by up to 50–60%, reducing water production and sewage treatment costs at centralised WWTPs. Other projects have focused on NBS and the water-food-energy nexus, as summarised in the Supplementary material tables. Hence, NBS include constructed wetlands, restored wetlands, coastal Mediterranean wetlands, green walls and green roofs. NBS, moreover, act as groundwater storage, water retention, water purification and improvement of environmental value.

The most frequent NBS are constructed wetlands, which can remove nutrients and organic components, including organic micropollutants and other emerging compounds. They can be designed for water sources with very different characteristics. In addition to treating water for a particular

doi: 10.2166/bgs.2020.932

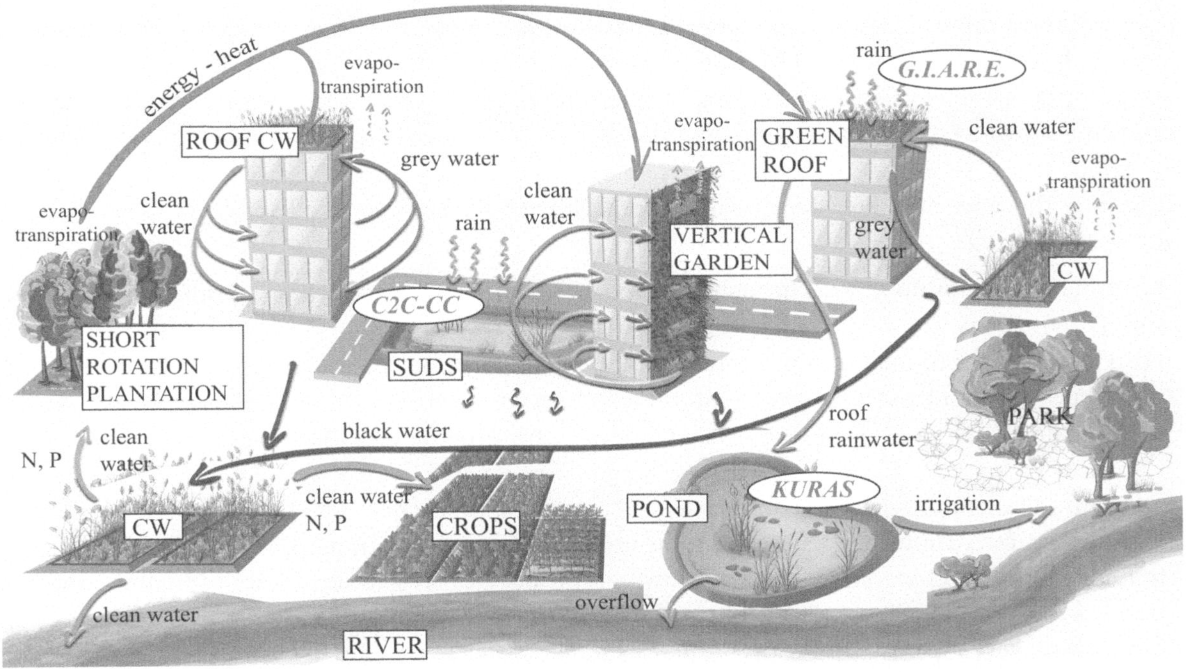

Figure 7 | Advisable scheme of decentralised and integrated, sustainable water management in an urban settlement with the contributions of the case studies C2C-CC, KURAS and G.I.A.R.E. highlighted (SUDS: sustainable drainage systems, CW: constructed wetlands) (adapted from Masi *et al.* 2018).

purpose, wetlands can be designed for water storage, infiltration and evapotranspiration, important functions of the urban water cycle. They also provide a series of additional benefits that grey infrastructure cannot, such as providing ecological niches within urban areas, or preferred recreation and educational areas. However, most of the research and technical development of treatment wetlands historically relates to decentralised treatment, normally away from urban areas. Thus, we still have only a small number of examples and limited data on the implementation of treatment wetlands in the urban environment. In spite of the potential of other NBS implemented in the urban area, such as green walls or SUDS, to purify water, the majority of the existing examples and publications deal only with the attenuation of the heat island effect or stormwater management, respectively. It is thus clear that in order to increase the implementation of NBS in the urban environment, further research and demonstration should more effectively combine different disciplines and needs in aligning with the holistic perspectives required by the water-food-energy nexus and taking into consideration the ecosystem services provided.

The implementation of NBS applications in urban areas is, at the same time, limited by some challenges. For instance, especially in densely built urban areas or protected historical city centres, the limited space available is a major drawback. Nevertheless, while this represents a present challenge, in the future, architecture and urban planning can be adapted to more easily accommodate NBS, which provide the widest possible range of benefits. In fact, NBS present a multifunctional capacity for resource recovery and pollution control, delivering multiple benefits in this issue, although it is worth noting that NBS for UWM clearly address other challenges, such as biodiversity enhancement and a more efficient management of the water-food-energy nexus, among others.

In the future, the reliance on NBS in sustainable water use is expected to increase. Given the still-increasing effects of climate change, it is necessary that in the future, planning for city infrastructure will be based on climate change mitigation, adaptation and resilience. The most common applications for NBS will be in parallel with integrated river management practices and re-establishment of wetlands. The developments towards more holistic concepts of resources flow management imply integrated, cross-sectoral systems and approaches. In the context of UWM, pollution control is shifting towards resource recovery. The traditional separation between water supply, wastewater and stormwater is challenged towards reusing properly treated wastewater and stormwater for purposes where potable water quality is not necessarily needed. NBS implementation can support this paradigm shift.

Based on the presented literature review, the NBS case studies and the discussion above, we conclude the following:

1. NBS help mitigate flood and drought impacts simultaneously supporting stormwater and water supply management.
2. NBS are essential to maintain the natural hydrologic regime despite development and partial sealing of surfaces, not least to keep the natural water cycle of evapotranspiration and rainfall, but also to mitigate urban heat island effects and allow the growth of urban green with local water resources.
3. NBS can efficiently purify very different water sources, greywater, rain water, sewer overflow or wastewater, for various purposes of further use, while generating numerous side benefits. Besides treating water, NBS can also retain stormwater, produce or irrigate for food production and save energy.
4. NBS create very promising new opportunities to use water more effectively and efficiently, enable urban farming or mitigate energy consumption. However, the urban water-food-energy nexus is still in a very early stage of development.
5. Ultimately, a wide application of NBS needs a systemic change from wanting to do things separately with various technologies towards learning to let nature take care of them in an integrated way that restores a close to natural local water balance and further important nature functions.

ACKNOWLEDGEMENTS

The study is carried out under the COST Action CA17133 Circular City (Implementing nature-based solutions for creating a resourceful circular city, http://www.circular-city.eu, duration 22 October 2018–21 October 2022). COST Actions are funded within the EU Horizon 2020 Programme. The authors are grateful for the support. All authors were funded by their respective affiliations. Gianluigi Buttiglieri acknowledges RYC-2014-16754, ICRA-TECH-2017 SGR 1318 and CTM2017-85385-C2-1-R.

SUPPLEMENTARY MATERIAL

The Supplementary Material for this paper is available at https://dx.doi.org/10.2166/bgs.2020.932.

REFERENCES

Anand, C. K. & Apul, D. S. 2014 Composting toilets as a sustainable alternative to urban sanitation – a review. *Waste Management* **34** (2), 329–343. doi:10.1016/j.wasman.2013.10.006.

Arabameri, A., Rezaei, K., Cerdà, A., Conoscenti, C. & Kalantari, Z. 2019 A comparison of statistical methods and multi-criteria decision making to map flood hazard susceptibility in Northern Iran. *Science of the Total Environment* **660**, 443–458. doi:10.1016/j.scitotenv.2019.01.021.

Arden, S. & Ma, X. 2018 Constructed wetlands for greywater recycle and reuse: a review. *Science of the Total Environment* **630**, 587–599. doi:10.1016/j.scitotenv.2018.02.218.

Barbosa, A. E., Fernandes, J. N. & David, L. M. 2012 Key issues for sustainable urban stormwater management. *Water Research* **46** (20), 6787–6798. doi: 10.1016/j.watres.2012.05.029.

Barceló, D. & Petrovic, M. 2008 *Emerging Contaminants From Industrial and Municipal Waste: Removal Technologies*. Springer, Berlin, Heidelberg, Germany. Available from: https://market.android.com/details?id=book-MF5GAAAAYAAJ

Bavor, H. J., Davies, C. M. & Sakadevan, K. 2001 Stormwater treatment: do constructed wetlands yield improved pollutant management performance over a detention pond system? *Water Science and Technology* **44** (11–12), 565–570.

Bawiec, A. 2019 Efficiency of nitrogen and phosphorus compounds removal in hydroponic wastewater treatment plant. *Environmental Technology* **40** (16), 2062–2072.

Bennett, G., Cassin, J. & Carroll, N. 2016 Natural infrastructure investment and implications for the nexus: a global overview. *Ecosystem Services* **17**, 293–297.

Bi, R., Zhou, C., Jia, Y., Wang, S., Li, P., Reichwaldt, E. S. & Liu, W. 2019 Giving waterbodies the treatment they need: a critical review of the application of constructed floating wetlands. *Journal of Environmental Management* **238**, 484–498.

Blue Green Solutions 2017 *A Systems Approach to Sustainable, Resilient and Cost-Efficient Urban Development, March 2017, Technical Report*. doi:10.13140/RG.2.2.30628.07046

Bockarjova, M. & Botzen, W. J. W. 2017 *Review of Economic Valuation of Nature-Based Solutions in Urban Areas. Technical Report, Naturvation Project*. Universiteit Utrecht.

Boogaard, F. C. 2015 *Stormwater Characteristic and New Testing Method for Certain Sustainable Urban Drainage System in the Netherlands*. Available from: https://doi.org/10.4233/uuid:d4cd80a8-41e2-49a5-8f41-f1efc1a0ef5d, http://citeseerx.ist.psu.edu/viewdoc/download?doi=10.1.1.909.8977&rep=rep1&type=pdf

Borisova, B. 2013 *Landscape Ecology and Landscape Planning*. Prof. Marin Drinov Academic Publishing House, Sofia, Bulgaria. ISBN 978-954-332-670-2. (in Bulgarian).

Bosch, P., Jongeneel, S., Rovers, V., Neumann, H.-M., Airaksinen, M. & Huovila, A. 2017 CITYkeys Indicators for Smart City Projects and Smart Cities. (CITYkeys project D1.4). Available from: http://nws.eurocities.eu/MediaShell/media/CITYkeysD14Indicatorsforsmartcityprojectsandsmartcities.pdf

Boyanova, K., Nedkov, S. & Burkhard, B. 2014 Quantification and mapping of flood regulating ecosystem services in different watersheds – case study in Bulgaria and Arizona, USA. In: *Thematic Cartography for the Society, Lecture Notes in Geoinformation and Cartography* (Bandrova, T., Konechy, M. & Zlatanova, S. eds). Springer, Cham, pp. 237–255.

Brown, R., Keath, N. & Wong, T. 2009 Urban water management in cities: historical, current and future regimes. *Water Science and Technology* **59** (5), 847–855. doi:10.2166/wst.2009.029.

Brunetti, G., Šimůnek, J. & Piro, P. 2016 A comprehensive numerical analysis of the hydraulic behavior of a permeable pavement. *Journal of Hydrology* **540**, 1146–1161. https://doi.org/10.1016/j.jhydrol.2016.07.030.

Brunetti, G., Šimůnek, J., Turco, M. & Piro, P. 2017 On the use of surrogate-based modeling for the numerical analysis of low impact development techniques. *Journal of Hydrology* **548**, 263–277. https://doi.org/10.1016/j.jhydrol.2017.03.013.

Bundesamt für Naturschutz 2017 *Urbane grüne Infrastruktur – Grundlage für attraktive und zukunftsfähige Städte. Hinweise für die kommunale Praxis. (Urban Green Infrastructure for Attractive and Sustainable Cities. Federal Agency for Nature Conservation)*. Bundesamt für Naturschutz, Bonn, Germany.

Burkhard, B., Kroll, F., Nedkov, S. & Müller, F. 2012 Mapping supply, demand and budgets of ecosystem services. *Ecological Indicators* **21**, 17–29.

Coast to Climate Change (C2-CCC) Project 2018 Available from: http://www.c2ccc.eu/. Last accessed 6 June 2019.

Carvalho, P. N., Basto, M. C. P., Almeida, C. M. R. & Brix, H. 2014 A review of plant–pharmaceutical interactions: from uptake and effects in crop plants to phytoremediation in constructed wetlands. *Environmental Science and Pollution Research* **21** (20), 11729–11763. doi:10.1007/s11356-014-2550-3.

Coffman, L. S. 1999 *Low Impact Development: A New Paradigm for Storm Water Management Mimicking and Restoring the Natural Hydrologic Regime*. Available from: https://www.waterboards.ca.gov/rwqcb2/water_issues/programs/stormwater/muni/nrdc/19%20lid%20design-%20a%20new%20paradigm.pdf

Cohen-Shacham, E., Walters, G., Janzen, C. & Maginnis, S. eds. 2016 *Nature-Based Solutions to Address Global Societal Challenges*. IUCN, Gland, Switzerland, p. xiii 97.

Comino, J. R., Keesstra, S. D. & Cerda, A. 2018 Connectivity assessment in Mediterranean vineyards using improved stock unearthing method, LiDAR and soil erosion field surveys. *Earth Surface Processes and Landforms* **43**, 2193–2206.

Dąbrowska, J., Bawiec, A., Pawęska, K., Kamińska, J. & Stodolak, R. 2017 Assessing the impact of wastewater effluent diversion on water quality. *Polish Journal of Environmental Studies* **26** (1), 9–16.

da Rocha, S. M., Almassy, D. & Pinter, L. 2017 *Social and Cultural Values and Impacts of Nature-Based Solutions and Natural Areas*. Naturvation project. Universiteit Utrecht, Utrecht, The Netherlands.

Depietri, Y. & McPhearson, T. 2017 Integrating the grey, green, and blue in cities: Nature-based solutions for climate change adaptation and risk reduction. In: *Nature-based Solutions to Climate Change Adaptation in Urban Areas* (Kabisch, N., Korn, H., Stadler, J. & Bonn, A., eds). Springer, pp. 91–109. http://dx.doi.org/10.1007/978-3-319-56091-5.

EPA-841-B-05-004 2015 *National Management Measures to Control Nonpoint Source Pollution From Urban Areas*. USEPA, Washington, DC, USA.

Eriksson, E., Baun, A., Scholes, L., Ledin, A., Ahlman, S., Revitt, M., Noutsopoulos, C. & Mikkelsen, P. S. 2007 Selected stormwater priority pollutants – a European perspective. *Science of the Total Environment* **383**, 41–51.

European Commission (EC) 2013 *Green Infrastructure (GI) – Enhancing Europe's Natural Capital*. COM 249 Final. Available from: http://ec.europa.eu/environment/nature/ecosystems/docs/green_infrastructures/1_EN_autre_document_travail_service_part1_ v2.pdf

European Commission (EC) 2015 *Nature-Based Solutions & Re-Naturing Cities*. Final Report of the Horizon 2020 Expert Group on 'Nature-Based Solutions and Re- Naturing Cities'. Directorate-General for Research and Innovation–Climate Action, Environment, Resource Efficiency and Raw Materials, p. 74. Available from: http://publications.europa.eu/resource/cellar/fb117980-d5aa-46df-8edc-af367cddc202.0001.04/DOC_1.

Faivre, N., Fritz, M., Freitas, T., de Boissezon, B. & Vandewoestijne, S. 2017 Nature-based solutions in the EU: innovating with nature to address social, economic and environmental challenges. *Environmental Research* **159**, 509–518.

Finger, D., Wüest, A. & Bossard, P. 2013 Effects of oligotrophication on primary production in peri-alpine lakes. *Water Resources Research* **49**, 1–11. doi:10.1029/wrcr.20355.

Fletcher, T. D., Andrieu, H. & Hamel, P. 2013 Understanding, management and modelling of urban hydrology and its consequences for receiving waters: a state of the art. *Advances in Water Resources* **51**, 261–279.

Fletcher, T. D., Shuster, W., Hunt, W. F., Ashley, R., Butler, D., Arthur, S., Trowsdale, S., Barraud, S., Semadeni-Davies, A., Bertrand-Krajewski, J. L. & Mikkelsen, P. S. 2015 SUDS, LID, BMPs, WSUD and more–the evolution and application of terminology surrounding urban drainage. *Urban Water Journal* **12** (7), 525–542.

Folke, C., Jansson, A., Larsson, J. & Costanza, R. 1997 Ecosystem appropriation of cities. *Ambio* **26** (3), 167–172.

Fonder, N. & Headley, T. 2013 The taxonomy of treatment wetlands: a proposed classification and nomenclature system. *Ecological Engineering* **51** (2), 203–211. doi:10.1016/j.ecoleng.2012.12.011.

Food and Agriculture Organization (FAO) 2018 *Nature-Based Solutions for Agricultural Water Management and Food Security*. FAO, Rome, Italy. Available from: http://www.fao.org/3/CA2525EN/ca2525en.pdf (accessed 14 October 2019).

Frantzeskaki, N. 2019 Seven lessons for planning nature-based solutions in cities. *Environmental Science & Policy*. **93** (3), 101–111. doi:10.1016/j.envsci.2018.12.033.

Gaines, J. M. 2016 Flooding: water potential. *Nature* **531**, S54–S55, https://doi.org/10.1038/531S54a.

Garcia-Rodríguez, A., Matamoros, V., Fontàs, C. & Salvadó, V. 2014 The ability of biologically based wastewater treatment systems to remove emerging organic contaminants–a review. *Environmental Science and Pollution Research* **21** (20), 11708–11728.

Garofalo, G., Palermo, S., Principato, F., Theodosiou, T. & Piro, P. 2016 The influence of hydrologic parameters on the hydraulic efficiency of an extensive green roof in Mediterranean area. *Water* **8** (44), 2–21. https://doi.org/10.3390/w8020044.

Gasperi, J., Zgheib, S., Cladiere, M., Rocher, V., Moilleron, R. & Chebbo, G. 2012 Priority pollutants in urban stormwater: part 2 – case of combined sewers. *Water Research* **46** (20), 6693–6703.

GIARE (Gestione Integrata Acqua Rinverdimento Energia) 2019 Available from: https:// www.giare.eu (accessed 10 October 2019).

Girard, L. F. & Nocca, F. 2019 Moving towards the circular economy/city model: which tools for operationalizing this model? *Sustainability* **11** (22), 6253.

Golden, H. E. & Hoghooghi, N. 2018 Green infrastructure and its catchment-scale effects: an emerging science. *Water* **5** (1), 1254. doi:10.1007/s11356-013-2448-5.

Gorito, A. M., Ribeiro, A. R., Almeida, C. M. R. & Silva, A. M. T. 2017 A review on the application of constructed wetlands for the removal of priority substances and contaminants of emerging concern listed in recently launched EU legislation. *Environmental Pollution* **227** (8), 428–443. doi:10.1016/j.envpol.2017.04.060.

Hansen, R., Frantzeskaki, N., McPhearson, T., Rall, E., Kabisch, N., Kaczorowska, A., Kain, J.-H., Artmann, M. & Pauleit, S. 2015 The uptake of the ecosystem services concept in planning discourses of European and American cities. *Ecosystem Serveruices* **12**, 228–246.

Hanson, H., Veerkamp, C., Nordin, A., Lazarova, T., Hedlung, K., Olsson, P. & Schipper, A. 2017 *Assessment of Biophysical and Ecological Services Provided by Urban Nature-Based Solutions. A Review*. Naturvation project. Universiteit Utrecht, Utrecht, The Netherlands.

Hill, G. B. & Baldwin, S. A. 2012 Vermicomposting toilets, an alternative to latrine style microbial composting toilets, prove far superior in mass reduction, pathogen destruction, compost quality, and operational cost. *Waste Management* **32** (10), 1811–1820. doi:10.1016/j.wasman.2012.04.023.

Hoyer, J., Dickhaut, W., Kronawitter, L. & Weber, B. 2011 *Water Sensitive Urban Design. Principles and Inspiration for Sustainable Stormwater Management in the City of the Future Manual. Elaborated in the Context of the Research Project SWITCH – Managing Water for the City of the Future*. HafenCity Universität, Hamburg, Germany.

HYDROUSA 2019 *Regenerative Water-Based Solutions*. Available from: https://www.hydrousa.org (accessed 11 October 2019).

Ilyas, H. & Masih, I. 2017 Intensification of constructed wetlands for land area reduction: a review. *Environmental Science and Pollution Research* **24** (13), 12081–12091. doi:10.1007/s11356-017-8740-z.

Ilyas, H. & van Hullebusch, E. D. 2019 Role of design and operational factors in the removal of pharmaceuticals by constructed wetlands. *Water* **11**, 2356. doi: 10.3390/w11112356.

Imfeld, G., Braeckevelt, M., Kuschk, P. & Richnow, H. H. 2009 Monitoring and assessing processes of organic chemicals removal in constructed wetlands. *Chemosphere* **74** (3), 349–362. doi:10.1016/j.chemosphere.2008.09.062.

Ingvertsen, S. T., Jensen, M. B. & Magid, J. 2011 A minimum data set of water quality parameters to assess and compare treatment efficiency of stormwater facilities. *Journal of Environmental Quality* **40** (5), 1488–1502.

Intergovernmental Panel on Climate Change (IPCC) 2015 Available from: https://www.ipcc.ch/

International Union for Conservation of Nature (IUCN) 2018 Available from: https://www.iucn.org/commissions/commission-ecosystem-management/our-work/nature-based-solutions (accessed 14 October 2019).

Kabisch, N., Frantzeskaki, N., Pauleit, S., Naumann, S., Davis, M., Artmann, M., Haase, D., Knapp, S., Korn, H., Stadler, J., Zaunberger, K. & Bonn, A. 2016 Nature-based solutions to climate change mitigation and adaptation in urban areas: perspectives on indicators, knowledge gaps, barriers, and opportunities for action. *Ecology and Society* **21**, 39.

Kalantari, Z., Ferreira, C. S. S., Keesstra, S. & Estouni, G. 2018 Ecosystem-based solutions for flood-drought risk mitigation in vulnerable urbanizing parts of East-Africa. *Current Opinion in Environmental Science & Health* **5**, 73–78.

Kazak, J. K. & van Hoof, J. 2018 Decision support systems for a sustainable management of the indoor and built environment. *Indoor and Built Environment* **27** (10), 1303–1306.

Kazak, J., Malczyk, J., Castro, D. G. & Szewrański, S. 2016 Carbon sequestration in forest valuation. *Real Estate Management and Valuation* **24** (1), 76–86.

Keesstra, S. D., Bouma, J., Wallinga, J., Tittonell, P., Smith, P., Cerdà, A., Montanarella, L., Quinton, J. N., Pachepsky, Y., van der Putten, W. H., Bardgett, R. D., Moolenaar, S., Mol, G., Jansen, B. & Fresco, L. O. 2016 The significance of soils and soil science towards realization of the United Nations Sustainable Development Goals. *Soil* **2**, 111–128.

Keesstra, S., Nunes, J., Novara, A., Finger, D., Avelar, D., Kalantari, Z. & Cerdà, A. 2018 The superior effect of nature based solutions in land management for enhancing ecosystem services. *Science of the Total Environment* **610**, 997–1009. doi: 10.1016/j.scitotenv.2017.08.077.

Kirchherr, J., Reike, D. & Hekkert, M. 2017 Conceptualizing the circular economy: an analysis of 114 definitions. *Resources, Conservation and Recycling* **127**, 221–232.

Kołecka, K., Obarska-Pempkowiak, H. & Gajewska, M. 2018 Treatment wetlands as the implementation of the circular economy. *Rocznik Ochrona Środowiska* **20**, 1350–1371.

Krauze, K. & Wagner, I. 2019 From classical water-ecosystem theories to nature-based solutions. contextualizing nature-based solutions for sustainable city. *Science of the Total Environment* **655**, 697–706.

Krzeminski, P., Tomei, M. C., Karaolia, P., Langenhoff, A., Almeida, C. M. R., Felis, E., Gritten, F., Andersen, H. R., Fernandes, T., Manaia, C. M., Rizzo, L. & Fatta-Kassinos, D. 2019 Performance of secondary wastewater treatment methods for the removal of contaminants of emerging concern implicated in crop uptake and antibiotic resistance spread: a review. *Science of the Total Environment* **648**, 1052–1081. doi:10.1016/j.scitotenv.2018.08.130.

KURAS 2019 Konzepte für urbane Regenwasserbewirtschaftung und Abwassersysteme (Concepts for Urban Rainwater Management and Sewage systems. Available from: www.watershare.eu (accessed 10 October 2019).

Lafortezza, R., Chen, J., van den Bosh, C. K. & Randrup, T. B. 2018 Nature-based solutions for resilient landscapes and cities. *Environmental Research* **165**, 431–441.

Lakatos, L., Żyromski, A. & Biniak-Pieróg, M. 2012 Possibility for modification of microclimate in orchards by using evaporative cooling irrigation. *Journal of Water and Land Development* **16** (1), 29–34.

Langergraber, G., Pucher, B., Simperler, L., Kisser, J., Katsou, E., Buehler, D., Garcia Mateo, M. C. & Atanasova, N. 2019 Implementing nature-based solutions for creating a resourceful circular city. *Blue-Green Systems* (in press).

Larsen, T. A. & Gujer, W. 1997 The concept of sustainable urban water management. *Water Science and Technology* **35** (9), 3–10.

Leigh, N. G. & Lee, H. 2019 Sustainable and resilient urban water systems: the role of decentralization and planning. *Sustainability* **11** (3), 918. 10.3390/su11030918.

Li, Y., Zhu, G., Ng, W. J. & Tan, S. K. 2014 A review on removing pharmaceutical contaminants from wastewater by constructed wetlands: design, performance and mechanism. *Science of the Total Environment* **468–469**, 908–932. doi:10.1016/j.scitotenv.2013.09.018.

Liquete, C., Udias, A., Conte, G., Grizzetti, B. & Masi, F. 2016 Integrated valuation of a nature-based solution for water pollution control. Highlighting hidden benefits. *Ecosystem Services* **22**, 392–401. doi:10.1016/j.ecoser.2016.09.011.

Loucks, D. P. & Van Beek, E. 2017 *Water Resource Systems Planning and Management –2017*. An Introduction to Methods, Models, and Applications. Springer, New York, USA.

Machado, A. I., Beretta, M., Fragoso, R. & Duarte, E. 2017 Overview of the state of the art of constructed wetlands for decentralized wastewater management in Brazil. *Journal of Environmental Management* **187** (2), 560–570. doi:10.1016/j.jenvman.2016.11.015.

Madrid, Y. & Zayas, Z. P. 2007 Water sampling: traditional methods and new approaches in water sampling strategy. *Trends in Analytical Chemistry* **26** (4), 293–299.

Maes, J. & Jacobs, S. 2017 Nature-based solutions for Europe's sustainable development. *Conservation Letters* **10** (1), 121–124.

Maes, J., Teller, A., Erhard, M., Liquete, C., Braat, L., Berry, P. & Paracchini, M. L. 2013 Mapping and assessment of ecosystems and their services. An analytical framework for ecosystem assessments under action. Technical report. *European Commission* **5**, 1–58.

Maiolo, M., Carini, M., Capano, G. & Piro, P. 2017 Synthetic sustainability index (SSI) based on life cycle assessment approach of low impact development in the Mediterranean area. *Cogent Engineering* **4** (1), 1410272. https://doi.org/10.1080/23311916.2017.1410272.

Makepeace, D. K., Smith, D. W. & Stanley, S. J. 1995 Urban stormwater quality: summary of contaminant data. *Critical Reviews in Environmental Science and Technology* **25** (2), 93–139.

Maksimovic, C., Mijic, A., Smith, K. M., Suter, I. & van Reeuwijk, M. 2017 *Blue Green Solutions. A Systems Approach to Sustainable, Resilient and Cost-Efficient Urban Development*. Technical report. https://doi.org/10.13140/RG.2.2.30628.07046.

Malaviya, P. & Singh, A. 2012 Constructed wetlands for management of urban stormwater runoff. *Critical Reviews in Environmental Science and Technology* **42** (20), 2153–2214.

Masi, F., Bresciani, R., Rizzo, A. & Conte, G. 2017 Constructed wetlands for combined sewer overflow treatment: ecosystem services at Gorla Maggiore, Italy. *Ecological Engineering* **98**, 427–438.

Masi, F., Rizzo, A. & Regelsberger, M. 2018 The role of constructed wetlands in a new circular economy, resource oriented, and ecosystem services paradigm. *Journal of Environmental Management* **216** (6), 275–284. doi:10.1016/j.jenvman.2017.11.086.

Matamoros, V., Jover, E. & Bayona, J. M. 2010 Occurrence and fate of benzothiazoles and benzotriazoles in constructed wetlands. *Water Science and Technology* **61** (1), 191–198. doi:10.2166/wst.2010.797.

Matzinger, A., Riechel, M., Remy, C., Schwarzmüller, H., Rouault, P., Schmidt, M., Offermann, M., Strehl, C., Nickel, D., Sieker, H., Pallasch, M., Köhler, M., Kaiser, D., Möller, C., Büter, B., Leßmann, D., von Tils, R., Säumel, I., Pille, L., Winkler, A., Bartel, H., Heise, S., Heinzmann, B., Joswig, K., Rehfeld-Klein, M. & Reichmann, B. 2017 *Zielorientierte Planung von Maßnahmen der Regenwasserbewirtschaftung - Ergebnisse des Projektes KURAS*. Berlin, Germany.

Natural Water Retention Measures 2015 *European Commission*. Available from: http://nwrm.eu/about-nwrm-project/about/project (accessed 17 October 2019).

Nikolaidis, N. P., Kolokotsa, D. & Banwart, S. A. 2017 Nature-based solutions: business. *Nature* **543** (7645), 315.

Nikolova, M. & Nedkov, S. 2018 Mapping and assessment of flood regulation ecosystem services in an urban environment. In: *Multifunctional Assessment Methods and the Role of Map Analyse–Using an Integrated Ecosystem Service Assessment Framework* (Potschin-Young, M. ed). Deliverable D4.8, EU Horizon 2020 ESMERALDA Project, Grant agreement No. 642007, pp. 53–67.

Palermo, S. A., Turco, M., Principato, F. & Piro, P. 2019 Hydrological effectiveness of an extensive green roof in Mediterranean climate. *Water* **11** (7), 1378. https://doi.org/10.3390/w11071378.

Parson, A., Bracken, L., Poeppl, R. E., Wainwright, J. & Keesstra, S. D. 2015 Introduction to special issue on connectivity in water and sediment dynamics. *Earth Surface Processes and Landforms* **40** (9), 1275–1277.

Paschke, A. 2003 Consideration of the physicochemical properties of sample matrices – an important step in sampling and sample preparation. *Trends in Analytical Chemistry* **22** (2), 78–89.

Piro, P., Carbone, M., De Simone, M., Maiolo, M., Bevilacqua, P. & Arcuri, N. 2018 Energy and hydraulic performance of a vegetated roof in sub-Mediterranean climate. *Sustainability* **10** (10), 3473. https://doi.org/10.3390/su10103473.

Piro, P., Turco, M., Palermo, S. A., Principato, F. & Brunetti, G. 2019 A comprehensive approach to stormwater management problems in the next generation drainage networks. In: *The Internet of Things for Smart Urban Ecosystems* (Cicirelli, F., Guerrieri, A., Mastroianni, C., Spezzano, G. & Vinci, A. eds). Internet of Things (Technology, Communications and Computing). Springer, Cham, Switzerland, pp. 275–304. https://doi.org/10.1007/978-3-319-96550-5_12.

Raymond, C. M., Berry, P., Breil, M., Nita, M. R., Kabisch, N., de Bel, M., Enzi, V., Frantzeskaki, N., Geneletti, D., Cardinaletti, M., Lovinger, L., Basnou, C., Monteiro, A., Robrecht, H., Sgrigna, G., Munari, L. & Calfapietra, C. 2017a *An Impact Evaluation Framework to Support Planning and Evaluation of Nature-Based Solutions Projects*. Report prepared by the EKLIPSE Expert Working Group on Nature-based Solutions to Promote Climate Resilience in Urban Areas. Centre for Ecology and Hydrology, Wallingford, UK.

Raymond, C. M., Frantzeskaki, N., Kabisch, N., Berry, P., Breil, M., Nita, M. R., Geneletti, D. & Calfapietra, C. 2017b A framework for assessing and implementing the co-benefits of nature-based solutions in urban areas. *Environmental Science & Policy* **77**, 15–24. doi:10.1016/j.envsci.2017.07.008.

Rizzo, A., Bresciani, R., Masi, F., Boano, F., Revelli, R. & Ridolfi, L. 2018 Flood reduction as an ecosystem service of constructed wetlands for combined sewer overflow. *Journal of Hydrology* **560**, 150–159.

Russo, A., Escobedo, F. J., Cirella, G. T. & Zerbe, S. 2017 Edible green infrastructure: an approach and review of provisioning ecosystem services and disservices in urban environments. *Agriculture, Ecosystem and Environment* **242**, 53–66.

Sustainable Development Goals Knowledge Platform 2018 *High Level Panel on Water*. Available from: https://sustainabledevelopment.un.org/HLPWater (accessed 24 November 2019).

Świąder, M. 2018 The implementation of the concept of environmental carrying capacity into spatial management of cities: a review. *Management of Environmental Quality* **29** (6), 1059–1074.

Świąder, M., Szewrański, S., Kazak, J. K., van Hoof, J., Lin, D., Wackernagel, M. & Alves, A. 2018 Application of ecological footprint accounting as a part of an integrated assessment of environmental carrying capacity: a case study of the footprint of food of a large city. *Resources* **7** (3), 52.

Szewrański, S., Chruściński, J., Kazak, J., Świąder, M., Tokarczyk-Dorociak, K. & Żmuda, R. 2018 Pluvial Flood Risk Assessment Tool (PFRA) for rainwater management and adaptation to climate change in newly urbanised areas. *Water* **10** (4), 386.

Szopińska, E., Kazak, J., Kempa, O. & Rubaszek, J. 2019 Spatial form of greenery in strategic environmental management in the context of urban adaptation to climate change. *Polish Journal of Environmental Studies* **28** (4), 2845–2856.

Talib, A. & Randhir, T. O. 2017 Managing emerging contaminants in watersheds: need for comprehensive, systems-based strategies. *Sustainability of Water Quality and Ecology* **9–10**, 1–8. doi:10.1016/j.swaqe.2016.05.002.

The Gorla Maggiore Water Park Project 2019 Available from: http:///naturvation.eu (accessed 4 June 2019).

United Nations 2015 *World Urbanization Prospects: The 2014 Revision. ST/ESA/SER.A/366.* UN Department of Economic and Social Affairs, Population Division, New York, USA.

Ursino, N. 2019 Dynamic models of socio-ecological systems predict catastrophic shifts following unsustainable development. *Science of the Total Environment* **654**, 890–894.

U.S. Environmental Protection Agency 2003 *Protecting Water Quality From Urban Runoff*. EPA 841-F-03-003. U.S. Environmental Protection Agency, Washington, DC, USA.

Verlicchi, P. & Zambello, E. 2014 How efficient are constructed wetlands in removing pharmaceuticals from untreated and treated urban wastewaters? a review. *Science of the Total Environment* **470–471**, 1281–1306. doi:10.1016/j.scitotenv.2013.10.085.

Vymazal, J. 2019 Is removal of organics and suspended solids in horizontal sub-surface flow constructed wetlands sustainable for twenty and more years? *Chemical Engineering Journal* **378**, 122117.

Wang, M., Zhang, D. Q., Dong, J. W. & Tan, S. K. 2017 Constructed wetlands for wastewater treatment in cold climate – a review. *Journal of Environmental Sciences* **57**, 293–311. doi:10.1016/j.jes.2016.12.019.

Winker, M., Gehrmann, S., Schramm, E., Zimmermann, M. & Rudolph-Cleff, A. 2019 Greening and cooling the city using novel urban water systems: a European perspective. In: *Approaches to Water Sensitive Urban Design* (Sharma, A. K., Begbie, D. & Gardner, T., eds). Potential, design, ecological health, urban greening, economics, policies, and community perceptions. Elsevier, Duxford, UK, pp. 431–455.

Wojciechowska, E., Gajewska, M., Żurkowska, N., Surówka, M. & Obarska-Pempkowiak, H. 2015 *Zrównoważone systemy gospodarowania wodą deszczową. (Sustainable Rainwater Management Systems)*. Gdansk University of Technology, Gdansk, Poland.

Wong, T. & Brown, R. 2009 The water sensitive city: principles and practice. *Water Science and Technology* **60** (3), 673–682. doi:10.2166/wst.2009.436.

WWAP (United Nations World Water Assessment Programme)/UN-Water 2018 *The United Nations World Water Development Report 2018: Nature-Based Solutions for Water*. UNESCO, Paris, France.

Zgheib, S., Moilleron, R. & Chebbo, G. 2012 Priority pollutants in urban stormwater: part 1 – case of separate storm sewers. *Water Research* **46**, 6683–6692.

Zhang, D., Gersberg, R. M., Ng, W. J. & Tan, S. K. 2014a Removal of pharmaceuticals and personal care products in aquatic plant-based systems: a review. *Environmental Pollution* **184**, 620–639. doi:10.1016/j.envpol.2013.09.009.

Zhang, D. Q., Jinadasa, K. B. S. N., Gersberg, R. M., Liu, Y., Ng, W. J. & Tan, S. K. 2014b Application of constructed wetlands for wastewater treatment in developing countries–a review of recent developments (2000–2013). *Journal of Environmental Management* **141**, 116–131. doi:10.1016/j.jenvman.2014.03.015.

Zhang, Y., Lv, T., Carvalho, P. N., Arias, C. A., Chen, Z. & Brix, H. 2016 Removal of the pharmaceuticals ibuprofen and iohexol by four wetland plant species in hydroponic culture: plant uptake and microbial degradation. *Environmental Science and Pollution Research International* **23** (3), 2890–2898. doi:10.1007/s11356-015-5552-x.

Zhang, Y., Liu, M., Zhou, M., Yang, H., Liang, L. & Gu, T. 2019 Microbial fuel cell hybrid systems for wastewater treatment and bioenergy production: synergistic effects, mechanisms and challenges. *Renewable and Sustainable Energy Reviews* **103**, 13–29. doi:10.1016/j.rser.2018.12.027.

Zhiyanski, M., Doncheva, S., Nedkov, S., Mondeshka, M., Yarlovska, N., Vassilev, V., Borisova, B., Chipev, N. & Gocheva, K. 2017 *Methodology for Assessment and Mapping of Urban Ecosystems Their State, and the Services That They Provide in Bulgaria*. Clorint. p. 82. ISBN.978-619-7379-03-7

Ziemiańska, M. & Kalbarczyk, R. 2018 Biometrics of tree-ring widths of (*Populus X canadensis* Moench) and their dependence on precipitation and air temperature in south-western Poland. *Wood Research* **63** (1), 57–75.

Zraunig, A., Estelrich, M., Gattringer, H., Kisser, J., Langergraber, G., Radtke, M., Rodriguez-Roda, I. & Buttiglieri, G. 2019 Long term decentralized greywater treatment for water reuse purposes in a tourist facility by Vertical Ecosystem. *Ecological Engineering* **138**, 138–147.

First received 1 July 2019; accepted in revised form 29 November 2019. Available online 17 January 2020

doi: 10.2166/bgs.2020.930

A review of nature-based solutions for resource recovery in cities

Johannes Kisser[a,*], Maria Wirth[a], Bart De Gusseme[b,c], Miriam Van Eekert[d,e], Grietje Zeeman[e], Andreas Schoenborn[f], Björn Vinnerås[g], David C. Finger[h], Sabina Kolbl Repinc[i], Tjaša Griessler Bulc[j], Aida Bani[k], Dolja Pavlova[l], Lucian C. Staicu[m], Merve Atasoy[n], Zeynep Cetecioglu[n], Marika Kokko[o], Berat Z. Haznedaroglu[p], Joachim Hansen[q], Darja Istenič[r], Eriona Canga[s], Simos Malamis[t], Margaret Camilleri-Fenech[u] and Luke Beesley[v]

[a] alchemia-nova GmbH, Institute for Innovative Phytochemistry & Closed Loop Processes, Vienna, Austria

[b] FARYS, Production & Transport TMVW, Gent, Belgium

[c] Center for Microbial Ecology and Technology, Ghent University, Gent, Belgium

[d] Environmental Technology, Wageningen University, Wageningen, The Netherlands

[e] LeAF BV, Bornse Weilanden, Wageningen, The Netherlands

[f] Institute of Natural Resources Science, Zurich University of Applied Science, Waedenswil, Switzerland

[g] Department of Energy & Technology, Swedish University of Agricultural Sciences, Uppsala, Sweden

[h] School of Science and Engineering, Reykjavik University, Reykjavik, Iceland

[i] Faculty of Civil and Geodetic Engineering, University of Ljubljana, Ljubljana, Slovenia

[j] Faculty of Health Sciences, University of Ljubljana, Ljubljana, Slovenia

[k] Agro-Environmental Department, Faculty of Agronomy and Environment, Agricultural University of Tirana, Tirana, Albania

[l] Department of Botany, Faculty of Biology, University of Sofia, Sofia, Bulgaria

[m] Faculty of Biology, University of Warsaw, Warsaw, Poland

[n] Department of Chemical Engineering, KTH Royal Institute of Technology, Stockholm, Sweden

[o] Faculty of Engineering and Natural Sciences, Tampere University, Tampere, Finland

[p] Institute of Environmental Sciences, Bogazici University, Bebek, Istanbul, Turkey

[q] Chair for Urban Water Management, University of Luxembourg, Luxembourg, Luxembourg

[r] Faculty of Health Sciences, University of Ljubljana, Ljubljana, Slovenia

[s] Department of Environment, Faculty of Urban Planning and Environmental Management, POLIS University, Tirana, Albania

[t] Department of Water Resources and Environmental Engineering, School of Civil Engineering, National Technical University of Athens, Zografou Campus, Athens, Greece

[u] Institute for Climate Change and Sustainable Development, University of Malta, Msida, Malta

[v] The James Hutton Institute, Craigiebuckler, Aberdeen, Scotland, UK

*Corresponding author. E-mail: jk@alchemia-nova.net

doi: 10.2166/bgs.2020.930

Abstract

Our modern cities are resource sinks designed on the current linear economic model which recovers very little of the original input. As the current model is not sustainable, a viable solution is to recover and reuse parts of the input. In this context, resource recovery using nature-based solutions (NBS) is gaining popularity worldwide. In this specific review, we focus on NBS as technologies that bring nature into cities and those that are derived from nature, using (micro)organisms as principal agents, provided they enable resource recovery. The findings presented in this work are based on an extensive literature review, as well as on original results of recent innovation projects across Europe. The case studies were collected by participants of the *COST Action Circular City*, which includes a portfolio of more than 92 projects. The present review article focuses on urban wastewater, industrial wastewater, municipal solid waste and gaseous effluents, the recoverable products (e.g., nutrients, nanoparticles, energy), as well as the implications of source separation and circularity by design. The analysis also includes assessment of the maturity of different technologies (technology readiness level) and the barriers that need to be overcome to accelerate the transition to resilient, self-sustainable cities of the future.

Key words: circular cities, energy, nature-based solutions, nutrients, resource recovery

ABBREVIATIONS/ACRONYMS

AD	Anaerobic digestion
ALE	Alginate-like exopolysaccharides
ATAD	Autothermal thermophilic aerobic digestion
Bio-W	Bio-waste
BIQ	Bio-intelligent quotient
BOD	Biological oxygen demand
BW	Blackwater
CDW	Construction and demolition waste
COD	Chemical oxygen demand
CSTR	Continuously stirred tank reactor
CW	Constructed wetland
DTM	Dry toilet matter
ESCO	Energy service companies
EWS	Evapotranspirative willow system
FGD	Flue gas desulphurisation
FW	Food waste
GHG	Greenhouse gas
GDP	Gross domestic product
GrW	Green waste
GW	Greywater
HRAP	High-rate algal ponds
K	Potassium
LCFA	Long-chain fatty acids
MBR	Membrane bioreactor
MBT	Mechanical biological treatment
MFCs	Microbial fuel cells
MSW	Municipal solid waste
N	Nitrogen
NBS	Nature-based solutions
OLAND	Oxygen-limited autotrophic nitrification/denitrification
P	Phosphorus
PBR	Photobioreactor
PCB	Polychlorinated biphenyl
PHA	Polyhydroxyalkanoates
PHB	Polyhydroxybutyrate
PPB	Purple phototrophic bacteria
R&D	Research and development
RO	Reverse osmosis
TRL	Technology readiness levels

TS	Total solids
TSS	Total suspended solids
UASB	Upflow anaerobic sludge blanket
VFAs	Volatile fatty acids
VFY	Vegetable, fruit and yard waste
VSS	Volatile suspended solids
WW	Wastewater
WWTP	Wastewater treatment plant
YW	Yellow water

INTRODUCTION

Cities are emerging as centres of human and economic capacity, with 54% of the global population living in cities and raising 85% of worldwide GDP (World Bank 2017). However, cities also accumulate or emit end-user resources and wastewater, functioning as resource sinks within the current linear economic model of 'take-make-dispose'. Urban populations consume 75% of natural resources, they are responsible for 50% of global waste and for 60–80% of overall greenhouse gas emissions (Ellen MacArthur Foundation 2017). Given the human and economic potential, the accumulation of resources and societal challenges of ecosystem degradation present in urban areas, the momentum is shifting towards recovery of these resources within the urban infrastructure.

Resource flows are generally considered 'waste', destined for final disposal as soon as they reach sewage systems, rubbish bins and exhaust pipes, although they include valuable resources such as nutrients (N, P, K), organics, water and metals. Each year, Europeans produce 3.6 Mt of N, 1.7 Mt of P and 1.3 Mt of K as part of human excrement. At the same time, Europe consumes 11 Mt of N, 2.9 Mt of P and 2.5 Mt of K of manufactured fertilisers (Fertilizers Europe 2017). The volumes of post-use material bear high potential. Therefore, the present study considers secondary resource streams, including urban wastewater, industrial wastewater, municipal solid waste and gaseous effluents, as well as the potential of source-separated waste(water) streams. Figure 1 showcases the urban water, nutrient, material and energy loops established by using and integrating nature-based solutions (NBS) in cities.

The present paper is a product of interdisciplinary cooperation among researchers from all 28 EU countries and 11 third countries within the EU-funded *COST Action Circular City*. Discussions among project members have produced a definition of nature-based solutions for the purpose of the COST Action, set out in Langergraber *et al.* (2019). As such, the present paper defines NBS as technologies that bring nature into cities and those that are derived from nature, using *organisms as principal agents* if they enable resource recovery and the restoration of ecosystem services in urban areas. The objective of this review is to provide a comprehensive overview of NBS applied and developed today to recover resources in cities, along current cutting-edge research and innovation, and to map out recoverable products as well as barriers, which represent the scope for further research. NBS can be applied to micro (household), meso (district) and macro (city and above) scales (Langergraber *et al.* 2019).

The findings are based on a literature review, as well as on the review of ongoing and recent research and innovation projects. These case studies were collected by participants of the *COST Action Circular City* with a portfolio of a total 92 research projects, as well as projects that partner researchers are aware of. Case studies specifically mentioned in the paper illustrate the diversity of applications and recoverable products. Based on discussions within the Action's working group on *resource recovery*, the present review paper looks at urban wastewater, industrial wastewater, municipal solid waste and gaseous effluents, as well as the implications of source separation of waste and end-of-pipe technologies versus circularity by design. Non-technical interrelated factors, which influence the applicability, selection and adoption of available technologies, such as legal frameworks, community awareness, acceptance and involvement, business and financing conditions, are not addressed here but are discussed in a separate review paper (Katsou *et al.* 2019).

doi: 10.2166/bgs.2020.930

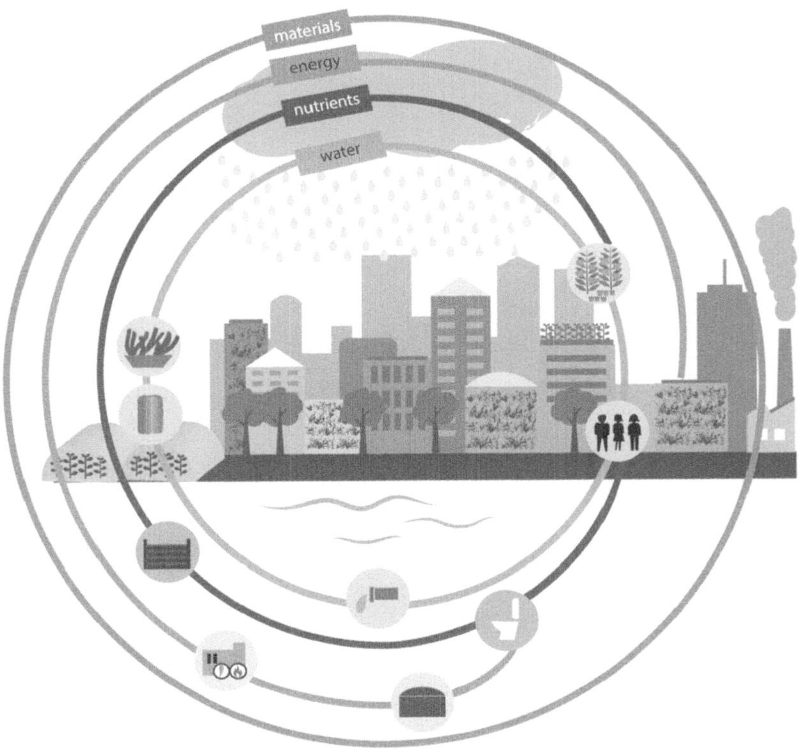

Figure 1 | Overview of urban water, nutrient, material and energy loops enabled using NBS within cities.

RESOURCE STREAMS AND OPPORTUNITIES FOR RECOVERY IN CITIES

The following section reviews different secondary resource streams found in cities, subdivided in urban wastewater, industrial waste and wastewater, municipal solid waste, gaseous effluents and source-separated waste. It provides an overview of technologies, projects and developments as well as barriers in relation to resource recovery with NBS.

Urban wastewater

Urban wastewater is defined as domestic wastewater or its mixture with industrial wastewater and/or runoff rainwater (European Commission 1991). The adequate treatment of urban wastewater is essential to protect human health and the environment. In Europe, cities largely collect and treat urban wastewater as a mixture of grey and blackwater, often also stormwater (combined sewer system). In Europe, more than 277 million people live in agglomerations bigger than 150,000 population equivalent (PE). They produce 41.5 million m^3 of wastewater per day. Currently, an annual 2.4% (1 billion m^3) of treated urban wastewater effluents are reused in the EU (European Commission 2018b), but this secondary resource stream bears significantly more resources to recover, including nutrients, organic carbon, lipids, biosolids and energy. The vast majority is still unexploited, but many of these can be recovered in cities using NBS. Table 1 provides an overview of projects deriving secondary resources and products from unsegregated urban wastewater, including reclaimed fertigation/irrigation water (water and nutrients), P-rich sludge, biopolymers, alginates, cellulose, construction material and energy (biogas, biofuel, electricity and heat). Information on the scale at which the technology is applied, the technology readiness level (TRL), region, project and project periods provide an indication as to the transferability of applied technologies.

Table 1 | Overview of resources that can be recovered from unsegregated urban wastewater, recovery technologies applied, recovered products, scale, TRL, region and project

Recoverable resource	Technologies applied	Products	Scale	TRL	Region	Project	Project period	Reference
Reclaimed water, energy and nutrients	Upflow anaerobic sludge blanket (UASB) + constructed wetlands + UV disinfection	Fertigation water; domestic non-potable water	Micro, meso	7	Lesvos Island, Greece	HYDROUSA	2018–2022	https://www.hydrousa.org/
	Combination with shred kitchen waste, liquid-solid separation, green walls, anaerobic membrane bioreactor (AnMBR), UV hygienisation	Fertigation water; biogas, fertiliser, domestic non-potable water	Micro	7	Austria, Spain	HOUSEFUL	2018–2022	https://houseful.eu/
	Hybrid constructed wetland, evapotranspirative willow system with zero discharge, algae-based technology	Fertigation water, woodchips for heat production	Micro, meso	7	Slovenia	GreenT (Slovenian Research Agency J2 – 8162 and Z2 – 6751)	2017–2020	http://www2.zf.uni-lj.si/si/component/content/article/32-raziskovanje-splosno/2489-zapiranje-snovnih-poti-pri-ciscenju-komunalnih-odpadnih-voda-z-zelenimi-tehnologijami-j2-8162
Nutrients	Adsorption columns and planted filters	Nutrients for irrigation water	Micro	3–4	Barcelona and Almería, Spain	INCOVER	2019–2021	https://incover-project.eu/technologies/nutrient-recovery
Organic carbon (carbohydrates)	Two-stage anaerobic-photosynthetic high-rate algal pond system	Biopolymers	Micro	6	Chiclana de la Frontera and Almería, Spain	INCOVER	2019–2021	https://incover-project.eu/technologies
	Two sequencing batch reactors (SBR): one for heterotrophic bacterial growth and the other for growth of autotrophic nitrifiers	Biopolymers (PHA) and P-rich sludge	Macro	6	Manresa, Spain	SMART-Plant	2016–2019	http://www.smart-plant.eu/
	Mixed microbial cultures, activated sludge at WWTP, bioprocess facilitating feast and famine conditions, biomass is fed with VFA-rich liquors, pure acetic and propionic acids	Biopolymers (PHA)	Meso	6–7	The Netherlands	Phario	2015–2019	http://phario.eu/
	Alginate extraction from granular excess sludge from 3 municipal Nereda®-plants and one industrial one	Alginates	Macro	6	WWTP Epe, Dinxperlo, Vroomshoop, The Netherlands	National Alginate Research Programme	2013–2019	https://www.royalhaskoningdhv.com/en-gb/news-room/news/water-authorities-working-hard-to-achieve-circular-economy/7123
	Alginate extraction from Nereda®-granular excess sludge	Kaumera Nereda® Gum (formerly: Neoalginate)	Macro	7–8	WWTP Zuthpen, The Netherlands	KAUMERA	2016–2018	https://kaumera.com/english/
Lipids	Two-stage aerobic/anaerobic reactor, *M. parvicella* bacterium accumulates FOG (fat, oil, grease), lipids extraction, subsequent esterification/transesterification	Biofuel	Micro	3–4	Luxembourg & France	WOW	2018–2021	https://www.cell-vation.com/wow-project

(*Continued*)

Table 1 | Continued

Recoverable resource	Technologies applied	Products	Scale	TRL	Region	Project	Project period	Reference
Energy, nutrients	Anaerobic biofilter for municipal wastewater treatment	Biogas	Meso, macro	7	Karmiel, Israel	SMART-Plant	2018–2020	http://www.smart-plant.eu/
	UASB for municipal wastewater	Biogas	Meso	7	Sweden	Pioneer-STP	2016–2019	https://www.kt.dtu.dk/english/research/prosys/projects/pioneer-stp
	Vacuum toilets and collection, AD, Fixed bed reactor, heat exchange, district heating	Biogas, fertiliser, thermal energy	Meso	7–8	Hamburg, Germany	Hamburg Water Cycle, Jenfelder Au	2011–2018	https://www.hamburgwatercycle.de/en/the-jenfelder-au-neighbourhood/the-hwc-in-the-jenfelder-au/
Energy	Horizontal subsurface CW with electrodes; oxidation of the organic matter generates electricity	Electricity	Micro	4–5	Spain, UK, Turkey	URBAN GreenUP	2017–2022	https://www.urbangreenup.eu/
Energy, salts	Microbial desalination combined with membrane treatment	Freshwater, treated wastewater	Micro, meso	5	Spain, Chile, Tunisia	MIDES	2016–2020	http://midesh2020.eu/
Nutrients, lipids, cellulose	Microbial conversion of nutrients to high-value compounds in a biorefinery approach	Ectoine, PHA, biogas, cellulose, construction materials	Meso, macro	6	Spain	DEEP PURPLE	2019–2023	https://deep-purple.eu/

DEEP PURPLE: Conversion of diluted mixed urban bio-wastes into sustainable materials and products in flexible purple photobiorefineries. GreenT: Closure of material pathways in urban wastewater treatment with green technologies. HYDROUSA: Demonstration of water loops with innovative regenerative business models for the Mediterranean region. INCOVER: Innovative eco-technologies for resource recovery from wastewater. MIDES: Microbial desalination for low energy drinking water. Pioneer-STP: The potential of innovative technologies to improve sustainability of sewage treatment plants. Run4Life: Large-scale nutrient recovery from domestic wastewater. SMART-Plant: Scale-up of low-carbon footprint material recovery techniques in existing wastewater treatment PLANTs. WOW: Wider business opportunities for raw materials from wastewater.

Technologies and products

As set out in Table 1, NBS for resource recovery from urban wastewater range from extensive technologies, such as constructed wetlands, evapotranspirative willow systems and algae ponds, to high-tech biological processes, such as rotating biological contactors, aerobic granulation (Nereda®) and anaerobic reactors. The wide range of recoverable products includes commonly derived products, such as biogas from primary and secondary sludge and reclaimed water for agricultural (crop irrigation or fertigation), industrial (cooling water), residential (sanitary flushing) and urban (park irrigation or even crop production) purposes as well as for groundwater recharge.

Combustible biomass of plants and microalgae can be converted to biogas and digestate for use as fertiliser through anaerobic digestion, bioethanol through sugar fermentation or ethylene reaction with steam (EUBIA 2019), biochar through pyrolysis, or processed for pulp-paper production or bioplastics. Bio-oil is produced by processing biomass under high temperature without oxygen and biohydrogen by steam reformation of bio-oils, dark and photofermentation of organic material as well as photolysis of water catalysed by specific microalgae species (Li *et al.* 2008). Algae biomass can also be used for feed production and extraction of high-value chemicals (Razzak *et al.* 2013; Passos *et al.* 2014; Wuang *et al.* 2016; Fermoso *et al.* 2019).

Constructed wetlands and nutrient-rich irrigation. Urban wastewater contains nitrogen and phosphorus which is usually not valorised within wastewater treatment plants (WWTPs). Although raw urban wastewater is a diluted effluent with low concentrations of nitrogen (30–70 mgN/L) and phosphorus (5–12 mgP/L), the large flows of generated wastewater carry significant quantities of nutrients. Constructed wetlands (CW) are the most common extensive NBS for nutrient recovery. They offer effective, reliable, robust and low-cost treatment of wastewater. Moreover, the nutrient content in the outflow can be adapted to the needs for crop fertigation. They can be integrated with other engineered solutions, such as anaerobic processes to meet strict water reuse regulations.

The EU-funded *HYDROUSA* project (Table 1) combines upflow anaerobic sludge blanket (UASB) with vertical constructed wetlands and UV disinfection to treat domestic sewage. The treated effluent is rich in nutrients, but has very low chemical oxygen demand (COD) and total suspended solids (TSS) levels, and is free of pathogens. It is used at the demonstration site to develop an agroforestry unit on the arid island of Lesvos, Greece, thereby reusing nutrients directly for agricultural purposes. The *HOUSEFUL* project (Table 1) also utilises domestic wastewater directly on site. It diverts the solids and liquids of the unsegregated household wastewater and treats the liquid fraction in green walls, hygenises it with UV radiation and reuses it for flushing toilets and irrigating food crops in greenhouses. The solids are co-digested together with the organic household waste in small biogas plants. The digested matter is converted to compost in a closed-vessel composting unit with in-built odour abatement (Bertino *et al.* 2018).

Numerous laboratory-scale experiments have been conducted introducing electrodes to (constructed) wetlands (e.g., iMETland or plant-e projects), generating electricity from the oxidation of the organic matter, but only a few pilot facilities have been attempted. The *URBAN GreenUP* project (Table 1) is piloting horizontal sub-surface flow (HSSF) wetlands, where electrodes and electrical connections through the filter bed stimulate the growth of an exoelectrogenic biofilm able to transfer the electrons generated by decomposition of organic matter. The *MIDES* project (Table 1) combines urban wastewater treatment and desalination by using microbial desalination processes to generate energy and run conventional reverse osmosis with the generated electricity.

The evapotranspirative willow system (EWS) (GreenT, see Table 1) treats wastewater and produces wood biomass. Mechanically pre-treated municipal wastewater flows into a waterproof bed filled with soil and planted with selected willow clones. In two research projects funded by the Slovenian Research Agency, willows in this system have been found to produce significantly more biomass

doi: 10.2166/bgs.2020.930

compared to control trees, namely 34–38 t DM/ha (Istenič *et al.* 2017, 2018). The treatment of wastewater produced by one person in a sub-Mediterranean climate requires 42 m^2 of EWS and produces 140–179 kg of wood biomass per year. Where available space allows the application of EWS, the wood biomass produced can be used for heating houses.

Microbial biotechnology. Anaerobic digestion is a popular treatment method for wastewater treatment sludge and enables recovery of energy (biogas, electricity, heat) and nutrients. Significant research is being conducted to enhance biogas and energy yields as well as valorisation of value-added products from side streams (intermediate products and valorisation of digestate). Among the projects mentioned in Table 1, *HYDROUSA, HOUSEFUL, SMART-Plant, Pioneer-STP* and *Hamburg Water Cycle/Jenfelder Au* are applying biomethane production using technologies such as common anaerobic digester, upflow anaerobic sludge blanket, anaerobic membrane bioreactor, anaerobic biofilter. Recently, biological production and harvesting of N_2O gas for energy recovery and reduction of high nitrogen loads in digestate centrate was performed by coupled aerobic-anoxic nitrous decomposition operation (CANDO). Combustion of N_2O with biogas increases energy yields and reduces the emission of the potent greenhouse gas (Weißbach *et al.* 2018).

Biofuel is usually produced from vegetable oils (soybean, canola, sunflower, palm and coconut oils) and animal fats, requiring large amounts of agricultural land. Urban wastewater can provide large quantities of alternative lipid feedstocks that help to meet the increasing demand for biofuel but do not compete with food production. Lipids, including oils, greases, fats and long-chain fatty acids are significant organic components of municipal wastewater, accounting for approximately 30–40% of the total COD of 120 g per PE and day, which means that about 18 kg per PE and year can be found in raw wastewater (Chipasa & Mędrzycka 2006). In the EU-funded WOW project (Table 1), lipids are accumulated by *Microthrix parvicella* bacteria and then processed to biofuel. The filamentous, selective lipid accumulator also has the ability to take up long-chain fatty acids, which can be used directly for the production of biofuel (Uwizeye *et al.* 2017).

The Nereda® process is a wastewater treatment technology, where activated sludge forms granules that have the ability to settle very fast. From these sludge granules, so-called 'alginate-like biopolymers' or 'alginate-like exopolysaccharides (ALE/Kaumera)' as a raw material can be obtained (Van der Roest *et al.* 2015). Aerobic granular sludge from the Nereda® process contains about 15–25% ALE that can be recovered. This material has the ability to bind strongly with water, can thicken and can also be used as a basis for coatings. The wastewater-derived alginate could be used for manifold applications, e.g., in the medical and food industries (Van der Roest *et al.* 2015). The neoalginate is already being recovered from granular sludge in three municipal WWTPs and one industrial plant in the Netherlands. The Zutphen WWTP produces 'Kaumera Nereda® Gum' (biopolymers), which can both retain and repel water. It is useful for a wide range of applications, e.g., in agriculture, to reduce leaching of fertilisers and enhance crop nutrient uptake, and in the concrete industry as a water-repellent coating for concrete floors (Waterschap Rijn en IJssel 2018).

Purple phototrophic bacteria (PPB) can convert organic matter from wastewater and from the organic fraction of municipal solid waste (MSW) into high-value compounds. Within the *DEEP PURPLE* project (Table 1), a PPB photobiorefinery is developed combining biomass, cellulose and biogas production in one single site. PPB uses near-infrared light as the main energy source, so they do not compete with other phototrophs such as microalgae or cyanobacteria (Madigan & Jung 2009).

Polyhydroxyalkanoates (PHAs) are bio-based and biodegradable thermoplastic polyesters. They are produced mostly from sugars or fats with pure culture fermentation. The *Phario* project (Table 1) is piloting a different approach, where secondary sludge from a municipal sewage treatment plant provides the functional biomass to produce PHA. Organic residues from the surrounding region were collected, fermented and successively fed to the sludge to produce a PHA-rich biomass with PHA content of 40–50% of the total volatile suspended solids (VSS). This PHA-rich biomass was acidified, dewatered by centrifugation and dried in a thermal dryer. The facility uses solvents such as butanol, which are reused (Bengtsson *et al.* 2017). The preliminary investigation was conducted in a pilot-scale facility in Brussels, using the full-scale secondary activated sludge from Bath WWTP (500,000 PE). The pilot has produced a biomass with PHA content of up to 0.47 g PHA/g VSS, which is above the considered profitability threshold (0.40 g PHA/g VSS) (Bengtsson *et al.* 2017). Each year, 2,000–2,500 t PHA can be produced from 2,500 t VSS of waste activated sludge generated in Bath WWTP. The results show that the harvested activated sludge could consistently yield PHA with high and controllable quality with fewer process elements, lower manufacturing costs and significantly lower environmental impact compared to currently available bioplastics.

Barriers

Reclaimed water and its treatment products can pose environmental, health and safety risks, which must be addressed during the development of resource recovery and water reuse systems. The products may contain pollutants and micropollutants like heavy metals, pharmaceuticals, personal care products, industrial chemicals, pesticides, microplastics, etc., which may enter the food chain through application to agricultural land. NBS can remove micropollutants often more effectively than conventional WWTPs (Guenther *et al.* 2002; Kabir *et al.* 2015; Gattringer *et al.* 2016; Balabanič *et al.* 2017), as conventional WWTPs are not designed to remove them. Due to their potential oestrogenic, mutagenic and carcinogenic activity (World Health Organization 2011), their removal and fate in NBS is of interest for the purpose of wastewater reuse and reclamation of other derived products.

An often-cited key barrier to the adoption of extensive technologies in densely populated areas (CW, algae systems and EWS) is the surface area requirement. However, microbial fuel cell technologies, active/passive aeration and innovative structural set-ups (e.g., vertECO® (Zraunig *et al.* 2019)) are already making CWs applicable even to cities. Furthermore, unutilised and underutilised urban spaces (including rooftops, facades, indoor spaces) could be used for nature-based urban wastewater treatment, resource reclamation and additional benefits, such as biodiversity, climate change mitigation and aesthetic/regenerative effects for the population. In order to facilitate the uptake of innovative rooftop and facade solutions, more demonstration projects are needed, to prove their functionality at relevant scales and a higher variety of contexts.

For research and non-research installations, the lack of standards, existing legal frameworks and lack of awareness of public administrative bodies make it very difficult to obtain building permits for these non-conventional systems. Authorities stick to existing laws and specific articles also for research purposes, as existing legal frameworks mostly do not include an exception for research. In the Netherlands, so-called Green Deals create a testing space for innovations for a certain timeframe (Rijksdienst voor Ondernemend Nederland 2019).

Further, the high number of derived end-products can result in competition between themselves, e.g., if lipids are extracted for biofuel production, the potential for biogas production is reduced. Practitioners and public entities often lack the know-how to identify the optimal biorefinery design and choice of secondary products in their individual cases. This calls for increased knowledge sharing for the available possibilities and selection parameters, including technical factors as well as economic factors (supply, demand, production costs, prices). Finally, some of the mentioned technologies have yet to mature in terms of technical readiness, the enabling legal and market framework, production costs and value chain as well as comprehensive impact assessments before they can be widely applied.

Industrial waste and industrial wastewater. Several raw and intermediate materials can be recovered from industrial waste streams using NBS. Studies at various scales exist for the recovery

doi: 10.2166/bgs.2020.930

of energy, carbon, nutrients, metals and chemicals from wastewater of pharmaceutical, chemical, food processing and metal industries (Mansouri *et al.* 2017; O'Dwyer *et al.* 2018; Song *et al.* 2018; Diaz-Elsayed *et al.* 2019). Table 2 gives an overview of recent and ongoing research projects recovering secondary resources and products from waste incineration as well as metal, dairy, food and pulp and paper industrial plants in cities.

Technologies and products

Phytomining. Phytomining is a 'green' alternative to opencast mining practices (Chaney *et al.* 2007) often causing environmental pollution. It is applied to recover a range of metals (Ni, Co, Au) but most often is used for Ni production in abandoned ferronickel mining sites (Osmani & Bani 2017; Osmani *et al.* 2018a, 2018b) and in naturally metalliferous soils (Li *et al.* 2003; Bani *et al.* 2015, 2018) because this raw material has gained high economic importance. The Ni-agromining chain consists of two stages: (1) the cultivation of hyperaccumulator plants to obtain sufficient aerial biomass with a high Ni concentration and (2) the transformation of the biomass to obtain valuable end-products. Both *in-situ* and *ex-situ* experiments were carried out in Albania, Spain, Austria and Greece, and Ni has been successfully recovered from bio-ores in pure form, as a mineral salt (ammonium nickel sulphate hexahydrate) or as eco-catalysts (Simonnot *et al.* 2018).

Using phytomining technology, the resulting ash is a real bio-ore, containing up to 20 weight percentages of Ni. It is possible to obtain different Ni compounds (e.g., Ni metal, Ni-based catalysts, Ni salts as ammonium nickel sulphate hexahydrate or oxides) by hydrometallurgical processes, where washing and refining processes are involved (Zhang *et al.* 2016; Houzelot *et al.* 2017, 2018). The cost of Ni is determined by the cost of the subsequent pyro- or hydrometallurgical processes. The production of Ni compounds such as ammonium nickel sulphate hexahydrate is a better alternative for Ni metal production, because of the higher price (97.50 EUR for 500 g with 98% purity, and 134 EUR for 25 g with 99.999% purity (Sigma-Aldrich 2018)).

Constructed wetlands. The food industry produces highly nutrient-rich solid waste and wastewater, which is a large untapped nutrient source. The *HIGHWET* project (Table 2) demonstrated constructed wetlands with reduced area successfully treating wastewater from food processing plants in Spain, Denmark and Belgium. The biomass can be processed to products mentioned above (in the section 'Urban wastewater').

Microbial biotechnology. Microbial biotechnology offers the advantage of using natural, high-affinity enzymes of different microorganisms that preferentially target the substrate of interest present in the industrial effluent (e.g., soluble selenium and other chemical elements, organic acids) to produce desirable end-products of industrial relevance (e.g., nanoparticles, biogas, biofuel). Examples of microbial technologies at microscale include Se nanoparticle recovery from waste streams of coal-fired power plants, bioelectrochemical metal recovery from metal and mining industry wastewaters, anaerobic digestion of dairy industry wastewaters for biogas production, fermentation of dairy or pulp and paper industry wastewaters for production of volatile fatty acids (VFAs) and/or hydrogen, and conversion of methanol in pulp industry wastewaters to VFAs with acetogenesis (Table 2).

Se is an essential micronutrient and a critical raw material with wide-range industrial utilisation (Hennebel *et al.* 2015). The current production of Se involves energy-intensive pyrometallurgical processing and smelting of Cu and Pb-ores, where it is recovered as an impurity. As a solution to its scarcity, Se could be recovered from industrial, secondary resources, such as effluents of flue gas desulphurisation (FGD), using cost-effective and environmental-friendly biotechnological approaches (Cordoba & Staicu 2018). Various bacterial groups can metabolise Se to generate cellular energy (i.e.,

Table 2 | Overview of recoverable resources from industrial waste and wastewater streams in cities, by secondary resource stream, recoverable resource, technologies applied, products, scale, TRL, region and project

Secondary resource stream	Recoverable resource	Technologies applied	Products	Scale	TRL	Region	Project	Project period	Reference
Bottom ash from incinerated MSW	Metals	Bioleaching	Enriched solution, Ga, Co, Mg, Cu, Zn, Al, Cr	Micro	3–4	Austria	GrecoMet	2016–2019	https://www.alchemia-nova.net/projects/grecomet/
Metal industry contaminated soil		Agromining	Nickel salt	Meso	6	Mediterranean climate	Life AgroMine	2016–2020	https://life-agromine.com/en/homepage/
Metal industry WW	Selenium, nanoparticles	Bioremediation coupled with resource recovery	Selenium nanoparticles	Micro	3–4	Temperate climate	Selenex	2018–2021	http://ddg.biol.uw.edu.pl/projects/staicu-sonata/
	Metals	Microbial fuel cell	Copper	Micro	3–4	The Netherlands, UK, Sweden, Finland, Spain, Luxembourg	BioElectroMET	2012–2016	http://www.bioelectromet.eu/
Dairy industry WW	Carbohydrates	Fermenter-bioaugmentation	VFAs	Micro	3–4	Sweden	EnVFAPro	2017–2018	https://www.kth.se/sv/ket/resource-recovery/envfapro-1.703273
	Energy	Anaerobic digestion	Methane	Micro	3–4	Denmark	ABWET	2015–2018	http://www.internationaldoctorate.unicas.it/abwet/
Food industry	Wastewater	Constructed wetland	Nutrient-rich biomass, clean water	Micro	5	Spain, Denmark, Belgium	HIGHWET	2013–2015	https://cordis.europa.eu/project/id/605445
Pulp industry craft mill foul condensate	Organic carbon	Acetogenesis (anaerobic digestion)	VFAs	Micro	3–4	Italy	ABWET	2015–2018	http://www.internationaldoctorate.unicas.it/abwet/
Pulp and paper industry WW	Carbohydrates	Dark fermentation	Hydrogen, VFAs	Micro	3–4	Italy	ABWET	2015–2018	http://www.internationaldoctorate.unicas.it/abwet/

ABWET: Advanced biological waste-to-energy technologies. EnVFAPro: Enhancement of volatile fatty acid production from dairy wastewater. GrecoMet: Green recovery of metals. HIGHWET: Performance and validation of HIGH-rate constructed WETlands. Selenex: Harvesting resources from industrial streams.

doi: 10.2166/bgs.2020.930

ATP) through anaerobic respiration, in parallel with the production of solid Se nanoparticles (Ni *et al.* 2015), as displayed in Figure 2.

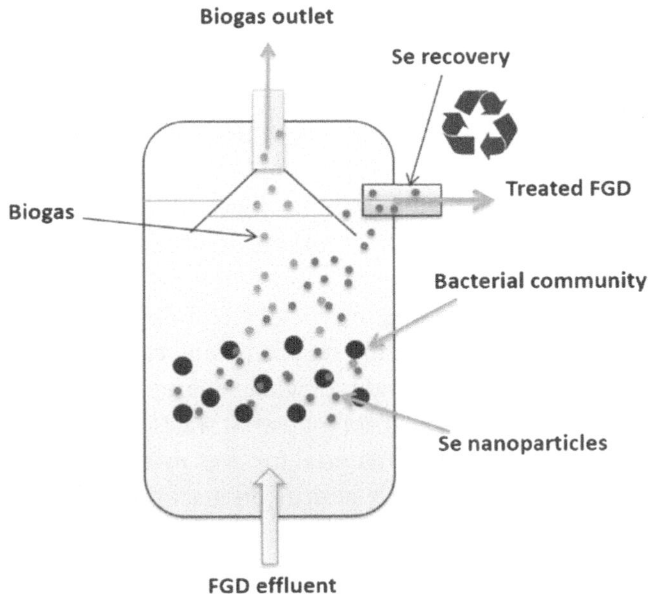

Figure 2 | Biological treatment and recovery of selenium using a biotechnological approach (modified from Cordoba & Staicu 2018).

Copper recovery from metallurgical waste and process streams using microbial fuel cells (MFCs) has been demonstrated. In MFCs, bacteria act as biocatalysts at the anode and generate current by oxidation of organic or inorganic substrates. The current can be used at the cathode to reduce, e.g., metal ions to solid metal species. Biological oxidation of either acetate (Rodenas Motos *et al.* 2015) or tetrathionate (Sulonen *et al.* 2018) has been coupled to Cu recovery in laboratory-scale MFCs. Furthermore, an MFC coupling acetate oxidation to Cu recovery was scaled-up to bio-anode and cathode surface areas of 835 cm^2 and 700 cm^2, respectively (Rodenas Motos *et al.* 2017).

Dairy industry wastewater contains high amounts of biodegradable carbon (Slavov 2017) and is a great source for the production of VFAs, which are valuable intermediate products of anaerobic digestion used in the conventional chemical industry. VFA on the market include formic, acetic, propionic, butyric, valeric and caproic acid. VFA have a wide range of applications, their recovery generates high production yield and releases less GHG emissions than biogas production (Atasoy *et al.* 2018). Bioaugmentation of the mixed cultures with pure *Clostridium aceticum* cultures proved to increase acetic acid production by 96 times, bioaugmentation with *C. butyricum* increased butyric acid production 120 times and *Propionibacterium acidipropionici* increased propionic acid production around five times compared to the control experiments. This case study (EnVFAPro project, Table 2) has shown that bio-based VFA production from waste streams can be environmentally friendly and economically feasible.

A pilot study (*HIGHWET* project, Table 2) at industrial food processing plants in Spain and Denmark tested the effect of effluent recirculation, aeration regime and different phosphorus adsorbent materials in a system that combines a hydrolytic up-flow sludge bed (HUSB) anaerobic digester as primary treatment, hybrid (vertical and horizontal flow (VF-HF)) constructed wetlands (CWs) and two different phosphorus adsorbent materials for treatment of the wastewater characterised by high nutrient loads. The project achieved a decrease of the required surface of conventional HFCWs and improved the final effluent quality in the aerated and non-aerated line, but the aerated VFCW was able to treat a four times higher loading rate with similar treatment efficiency than the non-aerated VFCW (Pascual *et al.* 2018).

Dairy wastewaters contain lipids that are hydrolysed into long-chain fatty acids (LCFA) that may be inhibitory to anaerobic microorganisms. Anaerobic conversion of LCFAs to methane was reported for the first time at 10 °C and 20 °C (with lipid content of >1%) in batch bottles, where the role of aceto-clastic methanogens from the genus *Methanosaeta* was highlighted (Singh *et al.* 2019). In the pulp industry, recovery of chemicals from black liquor results in the production of condensates that contain methanol up to 46 g/L. The pulp industry also produces thermomechanical pulping wastewaters that are released at high temperatures (50–80 °C). Conversion of methanol from condensates to VFA has been reported with an acetogenic culture in an up-flow anaerobic sludge bed reactor (Eregowda *et al.* 2018). Thermomechanical pulping wastewater, on the other hand, has been anaerobically converted to hydrogen at 70 °C with a culture dominated by *Thermoanaerobacterium* sp. (Dessì *et al.* 2018).

A variety of products can be recovered using microbial technologies, depending on the type of waste stream and desired recovered product, including metals (Wang *et al.* 2019), nanoparticles (Goethem *et al.* 2018), VFA (Zacharof & Lovitt 2014) and renewable energy carriers such as biogas and biofuel. Among a wide variety of recovered products from industrial waste streams, the described products are most promising with their potential as a raw material for post-processing. Most of the described technologies are still being developed at laboratory and/or pilot scales (e.g., bioelectrochemical systems, VFA production) (Chen *et al.* 2017; Garcia-Aguirre *et al.* 2017; Jankowska *et al.* 2017), except biogas production, which is established and implemented at full scale (Mauky *et al.* 2017; Martí-Herrero *et al.* 2019). In addition to biogas, bioplastic production has also been applied at pilot scale (Tamis *et al.* 2018). Mo *et al.* (2018) used food waste, fish waste and food processing waste to produce fish feed through biotransformation and solid-state fermentation.

Barriers

The phytomining techno-economic model should be customised to country-specific data reflecting differences in soil physicochemical properties in relation to the phytomining system implemented, Ni concentrations in the soils, hyperaccumulator yields and metal prices. The process efficiency and Ni salt purity are the main challenges of phytomining. Process parameters such as stirring speed or reaction time can significantly influence efficiency and they should be thoroughly investigated to assess their influence at each step of phytomining. One of the main limitations of energy recovery is the combustion temperature. Previous experiments demonstrate that combining energy recovery and utilisation of ashes for Ni recovery are compatible if the combustion temperature is low enough to avoid Ni losses through fly ash or other outputs. Preliminary calculations for Ni phytomining show promising results under the condition that heat released during incineration can be valorised close to the processing facility.

The main limitations are related to the complex matrix of industrial effluents, which often contain toxicants, that limit or prohibit bacterial growth. To overcome these hurdles, the recovery systems need to employ mixed microbial communities (as opposed to pure bacterial cultures). These mixed communities offer the advantage of protecting the microbial species of interest (e.g., metal respires or methanogens) against the toxic environment of the industrial effluent. However, on the other side, mixed communities also result in competition between various bacterial groups, some having better fitness and thermodynamics than the ones of interest for resource recovery (e.g., more thermodynamically adapted sulphate-reducing bacteria vs methanogens or metal oxidisers) (Hoelzle *et al.* 2014; Cetecioglu *et al.* 2019; Tang *et al.* 2019).

Another major challenge is the large variation in the composition and/or volume of the wastewaters, which may result from varying feed material composition, periodic operation or production intermissions due to, e.g., maintenance and cleaning. In addition, although it is a promising approach to recover bio-based products from industrial and municipal wastewater, there are still some technical

doi: 10.2166/bgs.2020.930

challenges such as product recovery after anaerobic digestion and purity of the recovered products (Puyol *et al.* 2017; Atasoy *et al.* 2018). Therefore, the microbial technologies should be able to cope with these changes, where mixed microbial communities again are more resilient than pure cultures. To reach full-scale adaptation, the microbial technologies should thus be able to handle high organic loading rates, regarding also high nitrogen and phosphorous concentrations and ensure sufficient wastewater treatment and resource recovery/product spectrum at varying wastewater conditions. By scaling up these systems, broad communication with stakeholders is crucial for preparing the market with new bio-products such as VFA.

Finally, many of the technologies that enable recovery of products other than energy are still in development and applied so far only at laboratory and pilot scales. The next step for these technologies will be scale-up to demo and flagship scales. However, already at this stage, the communication with public and private stakeholders is essential to prepare the market, including legislative and regulatory framework for the new bio-products.

Municipal solid waste. According to the European Commission (2019), municipal solid waste (MSW) constitutes about 10% of total waste generated in the EU. Although this figure may not seem too excessive at first glance, MSW is extensively prevalent and requires complex management linked to the mixed composition and multiple points of collection, which require various treatment methods. MSW includes waste streams from households and similar wastes from commerce, offices, public institutions and selected municipal services, excluding municipal sewage and construction and demolition waste (CDW). NBS applied to recover a wide range of intermediate and final products from mixed or biodegradable MSW include composting, anaerobic digestion and mechanical biological treatment (MBT). Research has also been conducted on bioleaching from mixed MSW incineration ash. Table 3 gives an overview of recent and ongoing research projects recovering secondary resources and products from MSW streams.

Technologies and products

Resource recovery from mixed MSW. Mechanical biological treatment (MBT) can enable recovery of ferrous metal, non-ferrous metal, plastic and glass from mixed MSW, but is mainly applied to stabilise MSW before landfilling. The biological steps include anaerobic digestion, composting and biodrying. Where recycling and recovery activities are low, it can improve environmental and economic performance (Trulli *et al.* 2018). However, MBT achieves only lower quality recyclates compared to those derived from recyclables from separate household collection, and mostly only metals are extracted. Digestate derived from mixed MSW is generally reported to be of lower quality than from separately collected organic waste, largely due to contamination with, e.g., glass and potentially toxic elements such as heavy metals (EPEM S.A. 2011). Biodrying is a partial composting stage, where the action of aerobic microbes rapidly heats and dries the waste. This process is used to produce a refuse-derived fuel that is dry and light for transport (Bogner *et al.* 2007).

Mixed (residual) MSW is often incinerated for electricity and heat production, and the incineration ash landfilled. The GRecoMet project (Table 3) (alchemia-nova 2019b) applied *Acidithiobacillus* bacteria (among other trials) to recover metals (finally selecting Cu, Cd and partially Co) from MSW incineration ash. The diffusely dispersed metals are brought into solution through microbial leaching, a process that efficiently extracts metals even from low-grade ores, such as MSW incineration ash (Chemiereport.at 2017). In the next steps, for enrichment of the dissolved metals, different NBS were tested, namely, enrichment in living and dead microalgae, rhizofiltration and sorption through peptides from microbial cells and waste biomass (biosorption). Hemp shives and sugar beet residues showed the highest sorption rates. Recovery of the pure metals from the metal-enriched

Table 3 | Overview of resources that can be recovered from different MSW streams, recovery technologies applied, recovered products, scale, TRL, region and project information

Secondary resource stream	Recoverable resource	Technologies applied	Products	Scale	TRL	Region	Project	Project period	Reference
Biodegradable fraction of MSW	Energy, nutrients	Separate collection at city level, centralised AD (digestion) and composting	Biogas, electricity and thermal heat, compost	Macro	8	Ljubljana, Slovenia	Centralised AT and composting at city level, e.g., RCERO	2007–2015	http://www.rcero-ljubljana.eu/upload/dokumenti/rcero_ljubljana_brusura_ang.pdf
		Separate collection at city level, centralised AD (digestion)	Methane for transportation, digestate (fertiliser)	Macro	8	Reykjavik, Iceland	Centralised methane recovery at city level CIRCLENERGY	2017–2018	https://www.carbonrecycling.is/circlenergy
	Nutrients, lipids, cellulose	Microbial conversion of nutrients to high-value compounds in a biorefinery approach	Ectoine, PHA, biogas, cellulose, construction materials	Meso, macro	6	Spain	DEEP PURPLE	2019–2023	https://deep-purple.eu/
	Organic carbon, energy, nutrients	Closed-vessel composting system with integrated plant biofilter	Biomass, odour removal, oxygen	Micro, meso	6	Austria, Greece, Spain	HYDROUSA, HOUSEFUL	2018–2022	https://www.alchemia-nova.net/projects/houseful/
Food waste and primary sludge	Carbohydrates	Acetogenesis (anaerobic digestion)	VFAs	Micro	3–4	Sweden	CarbonNextGen	2018–2020	https://resource-sip.se/projects/nasta-generations-koldioxidneutrala-avloppsreningsverk-carbonnextgen/
Food and garden waste + construction and demolition waste (CDW)	Green waste compost + crushed CDW material	Green waste compost and CDW are mixed 50:50	Improved soil-like substrate	Micro, meso	6	Scotland	The James Hutton Institute	2019–2021	https://www.hutton.ac.uk/staff/luke-beesley

DEEP PURPLE: Conversion of diluted mixed urban bio-wastes into sustainable materials and products in flexible purple photobiorefineries. HOUSEFUL: Innovative circular solutions and services for new business opportunities in the EU housing sector. HYDROUSA: Demonstration of water loops with innovative regenerative business models for the Mediterranean region. RCERO: Regional Waste Management Center of Ljubliana.

biomass was achieved through hydro- and pyrometallurgical pathways. The results suggested hydrometallurgical recovery directly from the leachate to be the most feasible option.

Resource recovery from the biodegradable fraction. If biodegradable municipal waste (garden and food waste from households, restaurants, supermarkets) is separated from other MSW at the source, it can be used as a carbon and nutrient source to produce several safe (uncontaminated) and valuable bio-based products (Atasoy *et al.* 2018). With 88 million tonnes of food waste produced in the EU every year (Kibler *et al.* 2018), this represents a waste stream with great potential for resource recovery. Composting and anaerobic digestion are commonly used processes.

Besides applying green waste compost (GWC) directly to fields and green spaces, it can also be mixed with deconstruction materials (CDW) to create a functional soil-like substrate (Table 3, *The James Hutton Institute*). CDW and GWC represent the mineral and organic parts of soil, respectively. In experiments growing ryegrass *Lolium perenne* and reed canary grass *Phalaris arundinacea*, a 50:50 volumetric ratio substrate yielded significantly greater biomass than other mixing ratios, and greater than that of the control soil (local topsoil). Such 'technical' soils and substrates can be produced from a range of urban wastes and, after physical, biological and chemical testing and verification, are envisaged as possible replacements to degraded or sealed soils in urban environments, creating bulk soils for the restoration of old capped landfill and mine site areas, and as alternative substrates for the growth of bioenergy crops (Nehls *et al.* 2015). Monitoring of leachates from such created substrates is required as materials such as CDW can contain high quantities of problematic components like gypsum, for example, which results in sulphate leaching.

Biomethane production and further heat and electricity production are common resource recovery technologies for kitchen waste (biodegradable fraction of MSW). Co-digestion of food waste with other waste, such as municipal wastewater (sludge) has been found to achieve a substantial increase of energy generation. Estimates of methane yields from various substrates can be found in the Methane Yield Database: online infrastructure and bioresource for methane yield data and related metadata (Murovec *et al.* 2015) (the database is freely accessible on the web page http://methane.fe.uni-lj.si/). The digestate is used as crop fertiliser (or soil conditioner) for microalgae cultivation, and in other cases further processed for biofuel and bioethanol production. As mentioned in the section 'Industrial waste and industrial wastewater', VFA are valuable intermediate products of anaerobic digestion. VFAs gained from food waste have also been processed to substrate for the production of biofuels, such as methane, hydrogen (e.g., Saadiah *et al.* 2017) and biofuel (Wang *et al.* 2019) as well as biopolymers such as polyhydroxyalkanoates (PHAs) (Raganati *et al.* 2014; Domingos *et al.* 2017). Physical, chemical and biological pre-treatment (via enzymes) methods exist to improve the degradation of cellulose and hemicellulose solubilisation (Strazzera *et al.* 2018), sugars' production and thus of VFAs (Braguglia *et al.* 2018). Atasoy *et al.* (2018) found that the organic fraction of MSW achieved the highest acidification and therefore highest yields after cheese whey and molasses (up to 40%).

Barriers

While composting and anaerobic digestion are well-established processes at mesoscale, the decentralised microscale for biogas production requires further research and development and is often confronted with legal barriers. Further, research to optimise anaerobic digestion is focused on improving biogas yield, while neglecting the quality of digestate (Logan & Visvanathan 2019). Logan & Visvanathan (2019) call for a shift from 'biogas optimisation' to 'integrated biogas-digestate optimisation'. Such an approach would consider potential value addition from digestate, which is generally not commercially exploited. Value addition with products for high-value markets is still in its infancy, with most attempts currently limited to laboratory or pilot scale.

Gaseous effluents. NBS can remove, contain and degrade gaseous contaminants into non-toxic or less toxic substances. These processes use the natural ability of plants to metabolise nutrients. They can also be enhanced by microbial and fungal communities colonising plant roots and above-ground organs of plants (e.g., Wood *et al.* 2006; Xu *et al.* 2011). Together, they can purify indoor and outdoor air from common pollutants including PMs (particulate matter), SO_2, NO_x, N_2O, O_3, VOCs (volatile organic compounds) (Wei *et al.* 2017), while also utilising CO_2 as a building block for plant biomass and releasing O_2. In doing so, these living biofilters can be used to transform polluted air into clean air and simultaneously produce plant biomass which can be processed into a range of secondary materials.

However, while terrestrial plants provide their aesthetic value and other co-benefits, the pollutant conversion and photosynthetic efficiency of microalgae are much higher. Microalgae (photosynthetic microorganisms, here including prokaryotic cyanobacteria and eukaryotes) can convert 10–20% of average solar energy in a mid-latitude region to biomass energy, versus 0.5% for the fastest-growing terrestrial plant, switchgrass (Li *et al.* 2008). Besides their high growth rate, microalgae can tolerate high CO_2 concentrations in gas streams; e.g., *Spirulina* sp., *Scenedesmus obliquus* and *Chlorella vulgaris* grow with up to 18% CO_2 (Morais & Costa 2007), allowing for high conversion efficiencies and enabling greater biomass harvests for further processing to biofuels including biogas, bio-oil, bio-hydrogen (Li *et al.* 2008). In addition to biofuel, which is a low-value, high-volume product, a number of high-value chemicals can be derived from microalgae and are already widely marketed, such as omega fatty acids and astaxanthin (Borowitzka 2013). The commercial cultivation of microalgae has rapidly increased over the last decades (Plaza *et al.* 2009).

NBS applied for resource recovery from gaseous effluents essentially include technologies using plants, plant-surrounding microorganisms as well as microalgae photobioreactors (PBRs) to store CO_2 and produce oxygen and biomass for further uses. These technologies are designed to purify ambient air, or by injecting gas directly into systems such as algae panels or tubes, or green walls. Table 4 gives an overview of recent and ongoing research projects recovering secondary resources and products from gaseous effluent streams in cities.

Technologies and products

The origin of plant-based air treatment goes back to the 1980s, when Wolverton *et al.* developed the first systems for NASA (Wolverton & McDonald 1983; Wolverton & McDonald-McCaleb 1986; Wolverton & Wolverton 1993). Within the last years, several plant-based air treatment systems have been developed at mesoscale, like Cloud Garden in the Netherlands (Cloud Garden 2019) and Green City Solutions in Germany (Green City Solutions 2019).

Green walls and microalgae structures are the most popular applications, usually applied with the foremost objective to purify ambient air in cities, i.e., bioremediation of indoor or outdoor air, binding or degrading various air pollutants. Especially, indoor air purification can have significant human health benefits as people in industrialised countries spend approximately 22 hours per day indoors. Air pollutants, which are generated indoors, e.g., VOCs, often accumulate due to limited ventilation (Pettit *et al.* 2018). Amid global warming, technologies such as active green walls, i.e., with active aeration, will gain importance due to their co-benefits of reducing indoor temperatures by 4–6 °C if close to an indoor wall (Fernandez-Cañero *et al.* 2012).

Outdoor structures with public visibility are typically designed to enhance the aesthetic value of urban spaces, such as green walls and the microalgae structures installed by EcoLogicStudio in the UK and other European countries. Green walls have been set up at all scales, from small indoor units to outdoor multi-storey facades, e.g., by Grünwand (Techmetall 2019) and the famous 'vertical forest', a high-rise apartment building in Milan designed by the architect Stefano Boeri, featuring 20,000 plants, 800 trees and over 100 different species. The vertical forest absorbs 40 tonnes of CO_2 and 1.5 tonnes of fine PM each year, generating 90 tonnes of oxygen per year (Bezemer 2017).

doi: 10.2166/bgs.2020.930

Table 4 | Overview of resources that can be recovered from different gaseous effluent streams found in cities, recovery technologies applied, recovered products, scale, TRL, region and project

Secondary resource stream	Recoverable resource	Technologies applied	Products	Scale	TRL	Region	Project	Project period	Reference
Vehicle exhaust gases, road traffic	CO_2-C, clean air	Glass tubular photobioreactors using algae	Combustible biomass, oxygen	Micro	7	Geneva, Switzerland	Culture Urbaine	2014	https://urbannext.net/culture-urbaine/
Outdoor air in urban spaces	Clean air	Plant-based green wall	Filtered air	Micro	5	EU	Green INSTRUCT	2016–2020	https://www.greeninstruct.eu/
		Large-scale green wall facade	Filtered air	Micro	8	Austria	Grünwand	2009–2013	https://gruenwand.com/
	CO_2-C, clean air	Mobile pods with tubular algae PBR structures	Oxygen, canopy area	Micro	3	Hungary	Chlorella Oxygen Pavilion	2012	Miklosi (2013)
		Curtain style vertically positioned algae reactor	Bioplastics, oxygen	Micro	6	United Kingdom	photo.Synthetica, EcoLogicStudio	Since 2018	https://www.photosynthetica.co.uk/
		Open algae tanks	Animal feed (protein), filtered air	Meso	7	Bangkok, Thailand	EnerGaia	Since 2009	https://energaia.com/
		Bio-wall type moss system	Filtered air	Micro	6	Germany	CityTree	Since 2015	https://greencitysolutions.de/en/
	CO_2-C, energy, clean air	Flat-panel photobioreactors (PBRs) using algae	Heat, biogas, oxygen	Micro	7	Hamburg, Germany	Building with Bio-Intelligent Quotient (BIQ)	2011–2013	https://www.buildup.eu/en/practices/cases/biq-house-first-algae-powered-building-world
		Wastewater treatment by open raceway algae ponds, anaerobic digestion, digestate dewatering, lipid extraction, biogas upgrading	Biofuel, biofertiliser, biomethane	Macro	6	El Torno Chiclana, Spain	All-Gas	2011–2016	http://www.all-gas.eu/en/
Industrial flue gas	CO_2-C, clean air	Vertically positioned plastic discs generating algae biofilms; continuous harvesting	Dry biomass, oxygen	Meso	5	Spain	ALGADISK	2012–2014	https://algen.eu/node/153
	CO_2-C, energy, clean air	Photobioreactor and photofermentation, anaerobic digestion of cyanobacteria residue	Bioplastic (polyhydroxybutric acid, PHB), biogas, nutrients for bacteria cultivation, fertiliser	Meso	6–7	Austria	CO_2USE	2012–2015	https://www.energy-innovation-austria.at/article/co2use-2/?lang=en
Indoor air	Clean air	Active hydroculture plant-based air treatment chambers	Filtered and humidified air	Micro	6	Denmark, UK, Switzerland, Spain	RECO$_2$ST	2018–2022	https://reco2st.eu/
Indoor air (households and other buildings), or flue gas from biogas CHP	CO_2-C, energy, clean air	Wall décor type algae biofilms; combination with biogas Combined Heat Power	Filtered air, biogas, electricity, heat	Micro	7	Germany	SOLAGA	Since 2015	https://www.solaga.de

ALGADISK: Novel algae-based solution for CO_2 capture and biomass production. Green INSTRUCT: Green INtegrated STRUCTural elements for retrofitting and new construction of buildings. ReCO2ST: Residential retrofit assessment platform and demonstrations for near zero energy and CO_2 emissions with optimum cost, health, comfort and environmental quality. SOLAGA: Living wall elements with algae.

As listed in Table 4, NBS can be used to derive a number of products from gaseous effluents. Plant-based technologies filter the air and convert CO_2 to biomass and O_2, producing opportunities for biomass processing to various mentioned products, while also improving ambient air quality. Low-value, high-volume products are mentioned in the section 'Urban wastewater'. High-value chemicals derived from microalgae include β-carotene, astaxanthin, docosahexaenoic acid, eicosahexaenoic acid, phycobilin pigments and algal extracts for use in cosmetics as well as polyunsaturated fatty acids, widespread 'superfoods' *Chlorella* and *Spirulina* (Borowitzka 2013), bioactive medicinal products, antioxidants, colouring agents and vitamins (Khan *et al.* 2018). Aromatic essential oils can be derived from plants used for phytoremediation. Processes such as steam distillation ensure that the oils are free from unwanted contaminants including heavy metals (Pandey & Souza-Alonso 2019). The following section describes case studies at micro, meso and macro level.

Micro. *RECO$_2$ST* (Table 4) is an EU-funded building renovation project aimed to achieve major energy savings through optimised refurbishment and integrated installation tools, including NBS, specifically, two biotechnical air treatment systems for purification, cooling and humidification of indoor air. The first is a mobile pot plant-based unit either as part of a retrofit or as a standalone unit. In the second system, ambient indoor air is treated by directing ventilation through a 'wintergarden'-like plant chamber. Both systems are hydroculture, with active aeration and automated sensors measuring air quality parameters. They can reduce PMs, VOCs, achieve stable indoor temperatures, rehydrate the air and enrich building aesthetics. As a result, overall quality of life, human health and productivity of the building inhabitants will be significantly improved. Current demo sites include apartment blocks in Frederikshavn (Denmark), London (UK), Vevey (Switzerland) and Cadiz (Spain). The ideal application is in office buildings, which are densely populated for many hours a day.

Meso. The *BIQ-building* (Table 4) in Hamburg, Germany, is the first algae-powered building in the world (IBA Hamburg GmbH 2013). Microalgae are bred in the glass facades, providing sufficient biomass to cover electricity and heat requirements of the whole building. Completed in 2013, *BIQ* is a five-storey, 15-apartment passive house designed by the Austrian architectural firm Splitter-Werk and funded by the Hamburg-based Climate Concept Foundation. The building features two types of photobioreactor (PBR) facades, where algae are grown for energy production as well as for controlling light and shade. The PBRs are filled with microalgae culture medium and supplemented with CO_2. Flue gas from a biogas-fuelled micro-CHP (combined heat and power) unit is injected into the PBRs. Circulated culture medium is collected at a central location within the building where recovered heat is drawn off by a heat exchanger and collected algal biomass is shipped to an off-site biogas unit. For infrastructural and legal reasons, biogas is not generated within the building. The PBR facades of *BIQ* generate 15 g total solids (TS) per m^2 per day across 200 m^2 (300-day indicator), yielding 2,600 m^3 methane and 6,000 kWh of net energy equivalent per day.

In the *CO$_2$USE* project (Table 4), cyanobacteria convert off-gas from an industrial production plant to biomass, which is further processed to bioplastic (PHB) as well as to biogas and digestate. The digestate is used to provide nutrients for bacteria cultivation and as common agricultural fertiliser. An ecological assessment showed that greenhouse gas emissions from PHB production can be up to 75% lower than for conventional polypropylene (BMVIT 2017).

Macro. In the EU-funded *All-Gas* project (Table 4), microalgae are cultivated in high-rate algal ponds (HRAP) with raceway design (with closed loop recirculation channels), filled with pre-treated urban wastewater. CO_2-containing flue gas from the biogas upgrading column is injected into the ponds and converted to algal biomass and further to secondary bio-products. An anaerobic digester converts the harvested algal biomass to biogas and digestate. Biofuel is gained through lipid extraction from dried digestate. The residue from lipid extraction is distributed as biofertiliser. The

doi: 10.2166/bgs.2020.930

total 4 ha site located at a municipal WWTP in Chiclana, Spain, generates around 400 tonnes of biomass per year.

Barriers

Challenges to comparison and further development of active botanical biofilters are the diverse experimental approaches assessing their performance, including different structural designs, different types and doses of pollutants as well as different time frames (Pettit *et al.* 2018).

Plant-based air purification systems are limited by their metabolic detoxifying capacity, thus requiring significant area compared to common purification systems. However, vertical structures enable greater plant density for floor space. Su & Lin (2015) found that, within an hour, a 6 m^2 indoor green wall could lower CO_2 concentrations from 2,000 to 800 ppm in a 39 m^3 room. In outdoor set-ups, the reduction rate is much smaller, but the aesthetic and stress-reduction potentials of greener cities argue for plant structures at larger scales. However, the maintenance required for healthy plants and their microbial populations remains a major drawback (Pettit *et al.* 2018). For plant systems, the use of invasive species poses a threat to sustainability and long-term feasibility (Pandey & Souza-Alonso 2019).

One side effect of plants, especially in cities, is their VOC emission. In that context, use of species from the genus of *Populus*, *Salix*, *Platanus* and others might be problematic. Isoprene emission from leaves of these species in summer months can increase formation of tropospheric ozone and other secondary pollutants in air (Sharkey *et al.* 2008). Consequently, a selection of plants with low VOC emissions themselves for plant-filter use is of great importance.

Another limitation is the diffusion of gaseous pollutants and associated removal inefficiencies, which can be mitigated by active airflow through plant substrate, e.g., active green walls (Pettit *et al.* 2018) or microalgae PBRs (Malinska & Zabochnicka-Swiatek 2010). On the other hand, high contaminant concentrations can inhibit plant and algae growth, i.e., their purifying activity. While microalgae growth is not limited by NO_x, SO_x concentrations above 400 ppm can lead to the formation of sulphurous acids and lower the pH. If the pH reaches below 4, the productivity of microalgae is reduced. This can be mitigated by applying NaOH to increase the pH (Malinska & Zabochnicka-Swiatek 2010). When microalgae (or plants) are harvested and processed for biofortification or fertiliser uses, careful analyses are necessary to exclude risk of contamination (Pandey & Souza-Alonso 2019). Closed PBRs overcome problems of external contamination (Malinska & Zabochnicka-Swiatek 2010). Regarding plant biofilters, it is suggested to use non-edible high-value crops for the treatment (Pandey & Souza-Alonso 2019).

Finally, a major challenge is that many secondary commercial products that can be derived from microalgae require further R&D to become profitable (Borowitzka 2013), such as PCB bioplastics (BMVIT 2017). The design of advanced PBRs, methods to enhance microalgae growth rates, the harvesting and drying methods, product synthesis and biomass pre-treatment are cited as crucial to improve cost-effectiveness of microalgae systems (Li *et al.* 2008; Malinska & Zabochnicka-Swiatek 2010; Khan *et al.* 2018). For mass microalgae production, flat plate and raceway PBRs are economically feasible, as opposed to horizontal tubular PBRs (Malinska & Zabochnicka-Swiatek 2010). Another factor for commercialisation is the highly disparate sizes of the markets for biofuels and high-value derivates, which may change in the light of current increased efforts to commercialise and develop new microalgae products (Borowitzka 2013).

SOURCE-SEPARATED WASTE

By implementing source separation solutions, domestic waste streams can be collected with higher nutrient levels and higher concentrations of organics (COD, BOD), for which clever sewage treatment and recovery technologies have been conceived. Such technologies minimise the release of toxic substances and protect natural freshwaters from eutrophication due to excess nutrient loadings (Finger *et al.* 2013). To obtain concentrated waste streams, dilution of solid and aqueous wastewater needs to be prevented. First of all, a separate sewer system with a sanitary and storm sewer can increase pollutant concentrations in wastewaters by around 85%, as calculated from typical German flow rates (Brombach *et al.* 2005).

Second, several options have been proposed for source separation at the household level of either urine (yellow water (YW)), using water-free urinals or source separation (NoMix) toilets and brown water (feces), or black water (BW). The latter waste stream combines urine and feces but in the selected case studies, dilution is avoided by means of vacuum toilets requiring low amounts of flushing water, and further separated vacuum transport. Another option is waterless dry toilets with or without urine separation. The collected dry toilet matter (DTM), depending on the type of toilet, can contain feces, urine, toilet paper and structural material. The sanitary wastewater from the laundry, kitchen, shower and bath is referred to as greywater (GW) and is separately collected as well. Finally, organic waste produced in cities can also be separated. We note the difference between kitchen waste (KW); bio-waste (Bio-W) referring to the combination of food waste and more general, the biodegradable fraction of catering waste; vegetable, fruit and yard waste (VFY), which is collected separately in several European cities; and green waste (GrW) collected in gardens and urban green spaces.

Coupling source separation to decentralised treatment/recovery of domestic wastewater, dry toilet matter and household waste (fractions) allows the recovery of valuable resources such as nutrients, organics, energy and water more efficiently. Table 5 gives an overview of recent and ongoing research projects recovering secondary resources and products from different source-separated waste and wastewater streams in cities.

Technologies and products

Micro. Sanitation 360 aims to produce fertiliser from human urine inside the toilet. The natural and fast enzymatic degradation of urea is chemically inhibited at pH 10 (Randall *et al.* 2016; Senecal & Vinnerås 2017; Simha *et al.* 2018). Thereafter, the water in the YW is evaporated and ventilated away leaving a fertiliser product with commercial-grade nutrient concentrations (>10% N, >1% P and >3% K). The decentralised inside-the-toilet approach to urine management allows large-scale implementation without major changes in the infrastructure, only requiring a new toilet and a drying bed. The first pilot systems have been implemented in single urine-diverting toilets in Sweden. A similar system has been implemented in the Autarky toilet developed at the Swiss Federal Institute of Aquatic Science and Technology (EAWAG) in Switzerland, where the YW is treated and used locally for fertiliser production (Larsen *et al.* 2015).

At the *Forum Chriesbach* office building in Duebendorf, Switzerland, a YW nutrient recovery system for 220 people has been in operation since 2012 (EAWAG 2019). YW is collected with waterless urinals and NoMix toilets and is directed to collection tanks in the basement. The urine is then nitrified in an aerated bioreactor (Etter *et al.* 2013), followed by a polishing step with activated carbon to eliminate pharmaceuticals and hormones. A vacuum distillation step reduces the liquid volume by 93% and eliminates pathogens. The product, a concentrated and processed urine-based fertiliser, contains all primary and secondary nutrients of the collected urine and is a fully approved fertiliser in Switzerland. It is produced and marketed as 'Aurin' by Vuna GmbH, a spin-off company of EAWAG (VUNA GmbH 2019). The main success factor was the determination of the EAWAG board to realise the new office building as a lighthouse project for integrated sustainable building practices, as well as the approval and support of the Swiss national authorities. The water and sanitation system was an important part of this broader context.

doi: 10.2166/bgs.2020.930

Table 5 | Overview of technologies applied to recover resources from source-separated urban waste(water), different secondary resource streams, recoverable resources, technologies applied, recovered products, scale, TRL, region and project

Secondary resource stream	Recoverable resource	Technologies applied	Products	Scale	TRL	Region	Project	Project period	Reference
Source-separated urban WW + kitchen waste	Reclaimed water, energy and nutrients	Vacuum collection, AD, OLAND, struvite precipitation, AD, heat exchange, district heating	Biogas, struvite fertiliser, thermal energy (heat)	Meso	8	Sneek, The Netherlands	Lemmerweg and Noorderhoek RUN4LIFE	2017–2021	http://run4life-project.eu/
		Vacuum toilets and collection, AD, struvite precipitation, AD in a membrane bioreactor, RO, heat exchange, district heating	Biogas, struvite fertiliser, heat, water reuse for industry	Meso	8	Ghent, Belgium	De Nieuwe Dokken RUN4LIFE	2017–2021	http://run4life-project.eu/
		Water-free urinals, vacuum toilets, AD, struvite precipitation	Biogas, struvite fertiliser, thermal energy (heat)	Meso	7–8	The Hague, The Netherlands	Rijkskantoor, Rijnstraat, NL	2017	https://www.saniwijzer.nl/projecten/rijkskantoor-rijnstraat-8/detail=94
Yellow water	Nutrients	Inside-the-toilet urine drying after chemical stabilisation	Dry fertiliser	Micro	7	Sweden	Urine dehydration technology for sanitation 2.0. Sanitation 360	2015–2018	https://www.slu.se/en/departments/energy-technology/projects/kretslopp/productive-on-site-sanitation-system/
		Water-free urinals, NoMix toilets, nitrification, activated carbon, distillation	Concentrated liquid fertiliser 'Aurin' (VUNA GmbH)	Micro	8	Duebendorf, Switzerland	VUNA – Nutrient Recovery from Urine	2010–2015	https://www.eawag.ch/en/department/eng/projects/vuna/
Grey water + dry toilet matter	Nutrients, organic carbon	Constructed wetland for greywater treatment; waterless dry toilets, composting and vermicomposting	Compost	Meso	8	Cressy, Geneva, Switzerland	Cooperative Equilibre @ Cressy	2011–2018	https://www.cooperative-equilibre.ch/projects/cressy/historique-de-limmeuble-de-cressy/
Grey water	Reclaimed water and nutrients	Green walls, vertical facade farming, vegetarian roof restaurants, aquaponics	Fertigation water	Meso, macro	7	Northern and central EU	EdiCitNet	2018–2023	https://cordis.europa.eu/project/rcn/216082/factsheet/en
Blackwater	Nutrients	Separate BW collection, centralised treatment with either ammonia sanitisation or AD with urea addition	Concentrated liquid fertiliser	Meso	7	Uddevalla[a], Västervik[b], Strängnäs[c], Örebro[d], Västerås[e] Sweden	Centralised BW treatment for >10 households	Implementation since [a,b]2013, [c]2014, [d]2015, [e]2018	https://pdfs.semanticscholar.org/f5dd/

EdiCitNet: Edible Cities Network Integrating Edible City Solutions for social resilient and sustainably productive cities. HOUSEFUL: Innovative circular solutions and services for new business opportunities in the EU housing sector.
[a,b,c,d,e]Dates refer to implementation of projects listed in the Region column.

Meso. The city of Sneek in the Netherlands has two areas with source separation systems: *Lemmerweg* (since 2005) and *Noorderhoek* (since 2010). BW is collected by means of vacuum collection (toilets) and transport systems that require about seven times less water (1 L per flush) than conventional sanitation. The developed sanitation concept (Zeeman *et al.* 2008) was first tested for several years with 32 houses at *Lemmerweg*, and subsequently applied for 232 households in *Noorderhoek*. The highly concentrated BW is mixed with ground KW and treated anaerobically in an upflow anaerobic sludge bed (UASB) reactor (Lettinga *et al.* 1981). A similar concept is now under construction for 550 houses in Amsterdam. The influent COD load is degraded, on average, for 70% resulting in a yearly biogas production of 10.5 $Nm^3/IE/a$ (Wit *et al.* 2018). Biogas energy is recovered as heat and used in a district heating system. Nitrogen is removed from the UASB effluent, using oxygen-limited autotrophic nitrification/denitrification (OLAND) (Vlaeminck *et al.* 2009). Phosphate is recovered as struvite and locally reused as fertiliser. GW is, together with the BW effluent, aerobically treated. GW has the highest temperature and energy potential to recover, and heat recovery through heat exchangers allows the transfer of most of the energy to the district heating system. A schematic representation of the

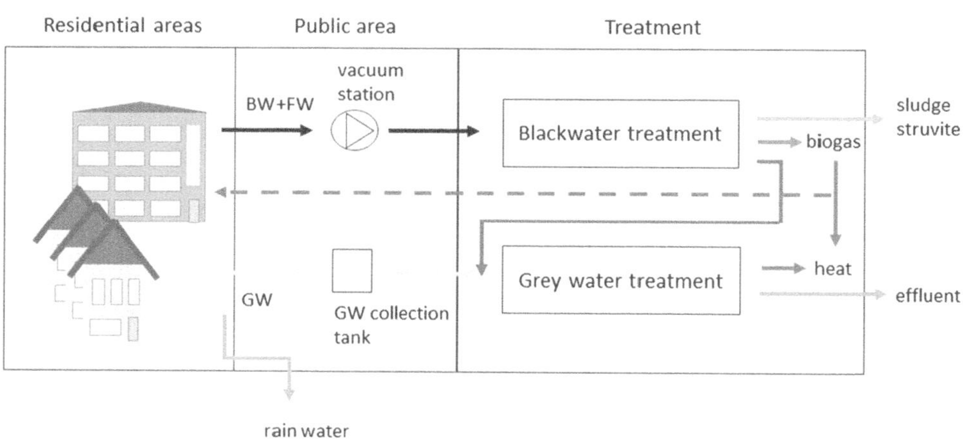

Figure 3 | Schematic representation of the projects in Sneek, The Netherlands (redrawn and adapted from Wit *et al.* 2018).

applied concept is given in Figure 3.

Similar examples are being set up throughout Europe. For example, the urban renewal project 'H+' in Helsingborg, Sweden consists of an old port and industrial areas, in which 320 apartments plus offices for 2,000 workers will have source separation systems. In order to reach its future sustainability goals, the city of Helsingborg has established cooperation between the municipal waste, energy and water companies. This Swedish eco-district is part of the *Run4Life* project, together with *De Nieuwe Dokken* district in Ghent, Belgium, and a new pilot at the previously mentioned *Lemmerweg*, and a pilot site in the industrial park of Porto do Molle (Vigo, Spain).

In the *De Nieuwe Dokken* project, the same decentralised treatment scheme will be applied as in the *Noorderhoek* project, similarly as depicted in Figure 3. The multi-step treatment concept is currently being set up for 400 households (1,265 IE), which will allow recovery of 1,600 kg/a struvite, to be used as slow-release fertiliser in the local green areas and urban farming projects, and up to 800 MWh_{th}/a through biogas utilisation and mostly GW excess heat recovery through heat exchangers (part of the NEREUS project, funded by the EU Interreg 2Seas program 2014–2020). In total, about one-third of the total heat demand of the urban area (2.1 GWh/a) can be provided by the decentralised treatment plant (Buysschaert *et al.* 2018). In contrast to the *Noorderhoek* projects, water reuse will be included after treatment of the GW in a membrane bioreactor (MBR), a cation exchange unit and reverse osmosis (RO) to remove pathogens, nutrients and hardness. In total, more than 30,000 m^3/a water will be reused as process water in a nearby factory. Furthermore, the excess heat of the factory will be recovered by coupling it to the district heating system, thus meeting the remaining two-thirds of the total heat demand. As such, this demonstration project couples recovery of energy and water, and the industrial activity and urban metabolism within a city.

Another example of source separation and recovery of resources at meso level is in the *Rijksgebouw* in The Hague, The Netherlands, which houses the Ministries of Foreign Affairs, Infrastructure and Water Management. The main incentive of the project at the Rijnlaan is to reduce water use and to recover nutrients and energy (Stichting 2017). In the basement of the building, source-separated YW and BW of ±6,000 office workers are treated. A part of the YW is collected in water-free urinals (approximately 200 L/day) and more than 95% of the phosphate is recovered as struvite. The struvite is applied in the facility as fertiliser in the enclosed garden of the office building. BW collected in vacuum toilets (3 m^3/day), and in the future combined with food waste (500 kg/day), is digested to produce biogas. The biogas is used in a central heating boiler, which is providing the energy for a hot water buffer tank to heat the building.

In the *Jenfelder Au* in Hamburg, Germany, the so-called *Hamburg Water Cycle®* is installed in a new neighbourhood for approximately 830 residential units. As in the other above-mentioned projects in Sneek, The Hague, Ghent and Helsingborg, BW (approximately 12 m^3/day) is collected via vacuum toilets and transported via a vacuum sewer to an anaerobic treatment system. In contrast with the other projects mentioned above in the Jenfelder Au, a mesophilic completely stirred tank reactor (CSTR) is implemented instead of a UASB reactor. Gas production in the Jenfelder Au is increased by adding external substrate from grease separators (maximum 30 m^3/day with approximately 6% dry matter). The digestate of the CSTR is expected to be used in agriculture after a post-treatment step (to be determined). Greywater will be treated via a fixed bed reactor in a first step. Further treatment processes will be examined in a test unit in order to determine the most effective one for different reuse purposes. The vacuum system has been in operation since 2017. The CSTR was opened in June 2019. The construction of the fixed bed reactor started at the end of 2019.

In Cressy (Geneva, Switzerland) the cooperative society 'Cooperative Equilibre' (CE) realised a three-storey/13-apartment building in 2011, which completely separates toilet waste from the water cycle. The toilet waste is collected with non-separating dry toilets, together with wood-chippings as structural material. The greywater is treated on-site in a constructed wetland. The DTM of each apartment is vermicomposted in the basement in a separate 1 m^3 container. Every 6–12 months, approximately 100 litres of pre-composted DTM are manually conveyed to a second composting step in the garden. After completion of the composting process (two years), the compost is used for fertilising trees and shrubs in the garden. Since 2011, CE has realised two more projects with a total of 103 apartments in Geneva following the idea of decentralised sanitation (including dry toilets) in an urban setting.

In areas with sensitive water recipients in Sweden, it is not allowed to apply (treated) BW as such into the environment. Lately, BW has been stored and sanitised in a large tank on a farm prior to reuse as fertiliser. The systems either sanitise the feces with ammonia sanitisation (addition of urea followed by >3 months storage) (Nordin *et al.* 2018), or a combination of biological (autothermal thermophilic aerobic digestion, ATAD) treatment followed by urea addition. ATAD increases the temperature allowing for less urea addition and shorter treatment time (Nordin & Vinnerås 2015). These centralised BW treatment systems have been set up in several municipalities in Sweden. Uddevalla has the largest number of connections with an annual treatment capacity of approximately 3,000 m^3, corresponding to 200–300 households. In total, over 1,000 households are covered with this type of system in over ten Swedish municipalities. In this way, nutrients are recycled for agriculture, with less transport of fertilisers and water.

Macro. Source-separated GrW is collected separately in many European countries and converted to energy and compost in large-scale centralised AT and composting facilities. In general, the waste treatment facilities apply thermophilic (dry) digestion and the digestate requires post-composting to stabilise. The compost is sold through commercial channels. This technology is used all over Europe to recover biogas/methane from separately collected organic waste and the mechanically separated fraction of organic waste from mixed waste. For example, *RCERO* in Ljubljana, Slovenia, is processing waste for around one-third of Slovenia (700,000 people), which amounts to around 150,000 tonnes of mixed municipal waste and 20,000 tonnes of separately collected food waste (Guardian 2019). The combined organic waste is treated in two-stage (thermophilic-mesophilic) plug-flow anaerobic reactors. Biogas is converted via a CHP to electricity and heat, which are both used on-site. The digestate (35,000 t/a) is dehydrated and further processed to produce 7,000 t/a compost.

Centralised methane recovery (Table 5) allows the use of the methane in biogas produced from bio-W, VFY and/or GrW at city level, in order to power, e.g., local transport. The city of Reykjavik, Iceland and its surrounding municipalities, home to about 150,000 inhabitants, collects all organic waste (60% biomass, 40% food waste) in a landfill (a bioreactor is being constructed). The biogas from the landfill has an exceptionally good quality with over 95% methane, which can be used directly in combustion engines. A pipeline from the landfill delivers the methane to gas stations for cars and trucks. About 2% of the personal cars run on methane, all of the city garbage trucks and some company trucks. In 2018, the construction of a modern biogas reactor was started, and it is estimated that methane production will triple, providing biofuel for up to 10% of the cars. A switch of the city buses from diesel to methane is under discussion. These actions are part of an ambitious climate action plan from the Icelandic government. The current government aims to ban registration of new gasoline and diesel buses by 2035 to become carbon neutral by 2040.

Barriers

Barriers for implementation of the above-mentioned concepts are related to the economy (of scale) and safety of operation. For example, the lack of sufficiently safe handling practices in the case of nutrient recovery from faeces or DTM poses a barrier. The removal of organic micropollutants and other contaminants such as microplastics, and hygienic safety of the recovered products are important as well. Recovery of COD and P is easier, since technologies are commercially available, but the recovery of N (and K) as a separate product is a problem. Only for streams with a very high N concentration, like urine (YW), are N-recovery technologies available, but during storage of urine a significant part of the nitrogen (about 50%) is potentially lost to the atmosphere due to premature hydrolysis to ammonia in, e.g., piping. Stripping of ammonia is technologically feasible but is currently not implemented because of the high energy and chemical demand. Another proven technique for N (and other nutrients) recovery from urine is nitrification (Udert & Wächter 2012), followed by a polishing step with activated carbon and vacuum distillation as applied for the urine collected in the EAWAG building. Another barrier for urine collection is the market availability of NoMix toilets. Several models have been removed from the market due to problems during use. Only dry toilets and water-free urinals are proven technologies. Recently, the new NoMix toilet 'Safe' was introduced and will probably be marketed at the beginning of 2020. A series of pilot projects in Switzerland and abroad are expected (EAWAG 2019).

When considering the reuse of products from domestic wastewater, for example as fertiliser in (urban) agriculture, the product quality is essential for environmental protection, as well as hygienic safety and user acceptance. In the case of dry toilets, the reuse of compost produced from DTM faces regulatory barriers if the compost is used beyond the owned plot of land. Another example of quality issues is contamination with microplastics since the implemented technologies for reuse of digestate and compost from municipal bio-waste do not completely remove microplastics (Weithmann *et al.* 2018). For example, Slovenia allows up to 0.5% (dry weight) of plastics that is larger than 2 mm in

doi: 10.2166/bgs.2020.930

compost and up to 2% (dry weight) of plastics in digestate. Technologies for achieving high-quality products, like membrane filtration, heating, AOP, activated carbon are available but will increase costs. The balance between risks and costs is to be established. In many European countries, the legal framework is currently limiting the possibilities of reusing products from wastewater in agriculture.

Scale is another factor of importance. The concept applied in Sneek, Ghent and Helsingborg is not suited for single or a few houses. Wit *et al.* (2018) calculate that this system becomes competitive with conventional sanitation (references: 30,000 and 100,000 inhabitants) at a scale of around 3,000 inhabitants (price of nutrient products is set at zero). In contrast, the concept applied in Cressy is limited to a maximum of three storeys, due to space constraints. Furthermore, the existing sanitation infrastructure represents an additional barrier since source separation sanitation requires new infrastructure. Most industrialised countries, however, are characterised by a high-density sewer network (with a very long lifetime) connected to municipal wastewater treatment plants; sewer and wastewater treatment plants have different lifetimes. According to Zeeman (2012), a gradual replacement is the only affordable way to introduce 'New Sanitation' at a larger scale, and the development of a transition strategy is required. Close cooperation between involved stakeholders, like that established in Sneek, Ghent and Helsingborg is crucial.

Another important aspect to convince the stakeholders and to remove the roadblocks for implementing new eco-technologies in an urban settlement is the integration of the local community and a sound business model, based on the development of new waste-based and circular value chains. Therefore, energy service companies (ESCO) can be set up to organise the technical maintenance and district services. For example, in the *De Nieuwe Dokken* project in Ghent, Belgium, the ESCO is a mixed private–public–citizen initiative in which the local inhabitants are represented, together with investors and public stakeholders such as the local water utility, FARYS. The local community will benefit directly from the revenues of the recovered products and the local district heating system. In the two projects at Sneek, The Netherlands, the conventional division of tasks was chosen in a cooperation between the housing cooperation – responsible for the indoor infrastructure (toilet and piping), the municipality – responsible for the outdoor infrastructure (vacuum station and sewer) and the water board – responsible for the treatment/recovery technologies. As in Ghent, the inhabitants pay the usual taxes and nothing more. A residents' satisfaction survey was done twice in the project in Sneek. Residents are predominantly satisfied with the system and consider it handy and hygienic, although some people had to get used to the vacuum toilet and kitchen grinder. The provided demonstration and the available information were highly appreciated (Wit *et al.* 2018).

DISCUSSION AND CONCLUSION

Resource recovery systems for urban residue streams comprise collection, transport, treatment/recovery and reuse. It is crucial to consider each step as, e.g., collection and transport will have an effect on applicable technologies for recovery and moreover on quality of products for reuse. When more dilution is allowed during collection and transport, the recovery technology becomes less (energy) efficient and more complex.

Common barriers

Considering barriers mentioned in the sections 'Urban wastewater', 'Industrial waste and industrial wastewater', 'Municipal solid waste' and 'Gaseous effluents', the realisation of the manifold potentials of NBS for circular cities faces a number of challenges. They can be divided into barriers related to

doi: 10.2166/bgs.2020.930

lack of awareness, current legislation, regulations and the organisation of urban infrastructures as well as technical barriers, raising the need for both further technical and social innovation.

Lack of awareness for proven capabilities of NBS

Even though they sometimes perform better than conventional grey technologies (e.g., see the section 'Urban wastewater'), NBS are de-prioritised. Despite many years of strong scientific track record, the capabilities of plants and microbes to convert nutrients into biomass, clean water and air, extraction of metals and other materials are not yet well known. Especially resource recovery projects using NBS in the narrow sense, i.e., as the European Commission understands them, plant-based systems delivering ecosystem services, are rare. Many NBS projects work to communicate their successes to policymakers and urban planners. Particularly large innovation and demonstration projects have the power to build trust and political willingness for broader implementation of NBS, and to overcome the lack of trust in NBS, even in industry. Capital expenditures for NBS are roughly on a par with conventional grey systems (depending on the type of systems compared), but NBS incur lower operational costs and offer additional benefits. Therefore, not only economic, but also environmental and social criteria can incentivise a shift from well-known grey technologies to NBS.

Legislative, regulatory and organisational barriers

The main barriers are related to uncertainties of new system financing (new business models, etc.) and the legislation in place (Houston (CSR Europe) *et al.* 2018). Further, once a resource becomes waste, a resource recovery effort often has to go through waste legislation, thus apply and fulfil all criteria for waste management. Even if applied in small scale, the efforts for application and documentation are similar to the requirements to run large recycling facilities. Also, current legislation does not always allow the direct reuse of secondary products. For example, the Netherlands currently faces ongoing discussions on how to deal with compost produced in the city. As local household compost is usually not tested and consequentially not approved, it cannot be easily applied across the city. Standards and legal frameworks need to adapt to scientific progress, but even research itself (not only implementation) is often already challenged by regulations, when there is no exemption clause in place for research purposes. A certain flexibility of administration processes and obligations could significantly stimulate wider implementation of NBS.

While the recovery of high-value products requires investments available only at macro- (and in some cases meso-) scale, micro- and mesoscale NBS bear the greatest potential for efficient nutrient and clean water recovery through direct reuse. As mentioned above, separate nutrient recovery with NBS is not feasible, but after pre-treatment, direct reuse of NBS-recovered secondary fertigation water and fertiliser/soil conditioner for urban agriculture can keep nutrients (and water) in highly efficient short cycles. This requires new management models in cooperation among municipalities and communities (neighbourhoods), innovation of the division of responsibilities among households/residents/local communities and municipalities (bottom up) coupled with spatial planning and simplification of applicable administrative hurdles (top down). The opportunities of resource recovery for value creation can be leveraged to incentivise decentralised ownership and maintenance.

Large advances have been achieved in reducing the area requirements of NBS, most notably constructed wetlands. Yet, availability of space in cities is still an often-cited barrier for functions such as CO_2 capture and wastewater treatment. There is need for more demo case studies and comparable evaluations that can provide standardised data on the ratio of surface area to functional efficiency for different technologies, climate and other conditions to support the planning process. Meanwhile, current planning and design models and tools used for centralised infrastructure approaches are not suitable for decentralised approaches and the integration of NBS into city-scapes. This calls for research

doi: 10.2166/bgs.2020.930

to identify the optimal scale, management scheme and logistics for existing specific conditions. Spatial planning innovations could facilitate the introduction of NBS to unutilised and underutilised infrastructures (rooftops, facades, indoor spaces). This could, in turn, allow for plant structures at larger scales, thus maximising the aesthetic and stress-reduction potentials of greener cities.

Further, the wide range of secondary end-products can lead to competition among different options. Therefore, there is a need for increased assessments of supply and demand factors, setting optimal configurations of NBS and blended green-blue-grey infrastructure and making the right choice of end-products.

Remaining technical barriers

Recovery of products other than energy is gaining momentum and there is a call for process optimisation towards product purity versus energy yield optimisation (e.g., 'integrated biogas-digestate optimisation' (Logan & Visvanathan 2019)). Many technologies that enable recovery of value-added products are still in development and applied so far only at laboratory and pilot scales. The next step for these technologies will be scale-up to demo and flagship scales, to prove the hygienic safety of waste(water)-derived products and to further diversify profitable high-value secondary products. Already at this stage, the communication with public and private stakeholders is essential to prepare the market including legislative and regulatory framework for the new bio-products. While NBS can provide essential functions for resource recovery, with significant additional benefits, further processing is usually required to achieve product purity required for commercialisation. Further, the toxicity of some raw industrial or municipal waste streams limits or even prohibits plant and microbial growth. In the field of source separation and decentralised applications, further research is needed to tackle the challenges mainly related to lack of economy (of scale) and safety of operation.

End-of-life management versus circularity by design

All these solutions look into recovery of secondary resources once they become waste. In this sense, they try to solve problems only at the end of the life cycle and have to take into account that many or most of the actual urban resource stream systems are not designed to be recovered. If you design a system from scratch with circular design in mind, the resource recovery would also be designed to happen with as little energy input as possible. The process can then even be designed to keep the resource value at the highest possible level (Bocken *et al.* 2016). By mixing resources with others, one has to apply more energy to again recover the value of one resource. In this sense, separation at or close to the source can be favourable for resource recovery purposes, although we should take into consideration the additional infrastructure needs and their associated grey energy (Larsen 2011). Direct metabolisation of organic nutrients from waste streams in agricultural systems can be one of the most favourable options (Capodaglio 2017).

The use of stored solar energy in organic resources for decentralised energy generation can also be a good approach, especially in combination with recovery processes. Since CO_2 is usually the last step in biomass energy systems, such a system can at best be climate-neutral. For more sustainable process designs, one additional aim can be the direct reuse of nutrients by building up biomass and simultaneously converting again CO_2 into biomass, as it is the building block of plants and many other phototrophic organisms. For reasonable carbon capture this biomass should then be either used in long-term storage systems like buildings, for furniture, etc., or should steadily be composted and integrated as increased soil carbon content. To take resource recovery with NBS to the next level, biorefinery approaches, also at a decentralised level, can be included. In this setting we have to look more into the feedstock quantity and quality of the different resource streams and the conversion to products. In the best case, the decentralised smaller biorefineries at the city level can pre-treat a certain organic residue stream and the conversion to bulk products can happen at a more centralised level (alchemia-nova 2019a). Appropriate logistics and a combination of zero km conversion of nutrients into food and exchange with the surrounding areas can be a good approach for cities. Cities can become 'major circular bioeconomy hubs' (European Commission 2018a).

The way forward

In this review paper, we identified projects, technologies and barriers for application of nature-based solutions for resource recovery in the framework of circular economy in cities. Our recommendations for further efforts are as follows:

- Replication of existing nature-based technologies for resource recovery in more cities and regional proof of concept for enabling further uptake.
- Upscaling existing and proven NBS resource recovery systems to bigger areas and for bigger settlements/regions/quarters.
- Raise the interest of investment schemes to fund more NBS cases.
- Demonstrate and stress the multifunctionality of NBS in new environments (e.g., industrial effluents or processes).
- Cooperate systematically with more actors along value chains and raise awareness.
- Share the know-how of NBS openly in developing or underprivileged countries.
- Using a value approach model as suitable means for a circular economy evaluation (e.g., value hill as tool (Achterberg & Fischer 2019)) together with other circular indicators (European Commission – Eurostat 2019).
- Comparison of direct reuse (metabolisation) of nutrients in agricultural systems vs technical recovery and shipping of nutrients back to the fields far away from the source.
- Comparing full cost accounting methods to direct nutrient conversion to agricultural produce with conventional farming systems.
- Awareness-raising for necessity of nutrient reuse from human systems and the hygienic quality of NBS.

ACKNOWLEDGEMENT

The collection of case studies happened primarily in the framework of the COST Action CA13177 Circular City ('Implementing nature based solutions for creating a resourceful circular city', http://www.circular-city.eu, duration 22 Oct 2018–21 Oct 2022) and is also deeply associated with the following projects: HYDROUSA (H2020-CIRC-2-2017, grant agreement no. 776643), HOUSEFUL (H2020-CIRC-1-2017, grant agreement no. 776708), ReCO2ST (H2020-EeB-05-2017, grant agreement no. 768576), Run4Life (H2020-CIRC-2-2016, grant agreement no. 730285).

REFERENCES

Achterberg, E. & Fischer, A. 2019 *3 Essential Steps to Financing Circular Business Models*. Check out These 3 Essential Steps to Learn How to Financing Your Own Circular Business Model! 2019. Circle Economy. Available from: http://www.circle-economy.com/financing-circular-business.

alchemia-nova 2019a *Austrian BioCycles*. Available from: https://www.alchemia-nova.net/projects/austrian-biocycles/.

alchemia-nova 2019b *Project GRecoMet*. Available from: https://www.alchemia-nova.net/projects/grecomet/.

Atasoy, M., Owusu-Agyeman, I., Plaza, E. & Cetecioglu, Z. 2018 Bio-based volatile fatty acid production and recovery from waste streams: current status and future challenges. *Bioresource Technology* **268**, 773–786. https://doi.org/10.1016/j.biortech.2018.07.042.

Balabanič, D., Filipič, M., Krivograd Klemenčič, A. & Žegura, B. 2017 Raw and biologically treated paper mill wastewater effluents and the recipient surface waters: cytotoxic and genotoxic activity and the presence of endocrine disrupting compounds. *The Science of the Total Environment* **574**, 78–89. https://doi.org/10.1016/j.scitotenv.2016.09.030.

Bani, A., Echevarria, G., Zhang, X., Benizri, E., Laubie, B., Morel, J. L. & Simonnot, M.-O. 2015 The effect of plant density in nickel-phytomining field experiments with *Alyssum murale* in Albania. *Australian Journal of Botany* **63**, 72–77. https://doi.org/10.1071/BT14285.

Bani, A., Echevarria, G., Pavlova, D., Shallari, S., Morel, J. L. & Sulce, S. 2018 Element case studies: nickel. In: *Agromining: Farming for Metals. Mineral Resource Reviews* (Van der Ent, A., Echevarria, G., Baker, A. & Morel, J., eds). Springer, Cham, Switzerland, pp. 221–232.

Bengtsson, S., Werker, A., Visser, C. & Korving, L. 2017 *PHARIO: Stepping Stone to A Sustainable Value Chain for PHA Bioplastic Using Municipal Activated Sludge.* Available from: http://edepot.wur.nl/413277.

Bertino, G., Menconi, F., Zraunig, A., Terzidis, E. & Kisser, J. 2018 *Innovative Circular Solutions and Services for New Buildings and Refurbishments.* WIT Press, Southampton, UK, pp. 83–91. https://doi.org/10.2495/ARC180081

Bezemer, M. 2017 Phytoremediation: How Plants Help Restore Balance to Our Environment. *Into Green* (blog). December 7, 2017. Available from: https://intogreen.eu/phytoremediation-how-plants-help-restore-balance-to-our-environment/.

BMVIT 2017 CO2USE. *Energy-Innovation-Austria* (blog). 2017. Available from: https://www.energy-innovation-austria.at/article/co2use-2/?lang=en.

Bocken, N. M. P., de Pauw, I., Bakker, C. & van der Grinten, B. 2016 Product design and business model strategies for a circular economy. *Journal of Industrial and Production Engineering* **33** (5), 308–320. https://doi.org/10.1080/21681015.2016.1172124.

Bogner, J., Abdelrafie Ahmed, M., Diaz, C., Faaij, A., Gao, Q., Hashimoto, S., Mareckova, K., Pipatti, R. & Zhang, T. 2007 Waste management. In: *Climate Change 2007: Mitigation. Contribution of Working Group III to the Fourth Assessment Report of the Intergovernmental Panel on Climate Change* (Metz, B., Davidson, O. R., Bosch, P. R., Dave, R. & Meyer, L. A., eds). Cambridge University Press, New York, USA, pp. 585–618.

Borowitzka, M. A. 2013 High-value products from microalgae – their development and commercialisation. *Journal of Applied Phycology* **25** (3), 743–756. https://doi.org/10.1007/s10811-013-9983-9.

Braguglia, C. M., Gallipoli, A., Gianico, A. & Pagliaccia, P. 2018 Anaerobic bioconversion of food waste into energy: a critical review. *Bioresource Technology, Bioconversion of Food Wastes* **248**, 37–56. https://doi.org/10.1016/j.biortech.2017.06.145.

Brombach, H., Weiss, G. & Fuchs, S. 2005 A new database on urban runoff pollution: comparison of separate and combined sewer systems. *Water Science and Technology* **51** (2), 119–128.

Buysschaert, B., Vermijs, L., Naka, A., Boon, N. & De Gusseme, B. 2018 Online flow cytometric monitoring of microbial water quality in a full-scale water treatment plant. *Npj Clean Water* **1** (1), 16. https://doi.org/10.1038/s41545-018-0017-7.

Capodaglio, A. G. 2017 Integrated, decentralized wastewater management for resource recovery in rural and peri-urban areas. *Resources* **6** (2), 22. https://doi.org/10.3390/resources6020022.

Cetecioglu, Z., Dolfing, J., Taylor, J., Purdy, K. J. & Eyice, O. 2019 COD/sulfate ratio does not affect the methane yield and microbial diversity in anaerobic digesters. *Water Research* **155**, 444–454. https://doi.org/10.1016/j.watres.2019.02.038.

Chaney, R. L., Angle, J. S., Broadhurst, C. L., Peters, C. A., Tappero, R. V. & Sparks, D. L. 2007 Improved understanding of hyperaccumulation yields commercial phytoextraction and phytomining technologies. *Journal of Environmental Quality* **36** (5), 1429–1443. https://doi.org/10.2134/jeq2006.0514.

Chemiereport.at 2017 Metall Aus Müll-Asche. *Österreichs Magazin Für Wirtschaft, Technik Und Forschung* **2017**. Available from: https://www.chemiereport.at/epaper/201705/epaper/ausgabe.pdf.

Chen, X., Zhou, H., Zuo, K., Zhou, Y., Wang, Q., Sun, D., Gao, Y., Liang, P., Zhang, X. & Ren, Z. J. 2017 Self-sustaining advanced wastewater purification and simultaneous in situ nutrient recovery in a novel bioelectrochemical system. *Chemical Engineering Journal* **330**, 692–697. https://doi.org/10.1016/j.cej.2017.07.130.

Chipasa, K. B. & Mędrzycka, K. 2006 Behavior of lipids in biological wastewater treatment processes. *Journal of Industrial Microbiology and Biotechnology* **33** (8), 635–645. https://doi.org/10.1007/s10295-006-0099-y.

Cloud Garden 2019 *Cloud Garden.* Available from: https://www.cloudgarden.nl/en/.

Cordoba, P. & Staicu, L. C. 2018 Flue gas desulfurization effluents: an unexploited selenium resource. *Fuel* **223**, 268–276.

Dessì, P., Porca, E., Lakaniemi, A.-M., Collins, G. & Lens, P. N. L. 2018 Temperature control as key factor for optimal biohydrogen production from thermomechanical pulping wastewater. *Biochemical Engineering Journal* **137**, 214–221. https://doi.org/10.1016/j.bej.2018.05.027.

Diaz-Elsayed, N., Rezaei, N., Guo, T., Mohebbi, S. & Zhang, Q. 2019 Wastewater-based resource recovery technologies across scale: a review. *Resources, Conservation and Recycling* **145**, 94–112.

Domingos, J. M. B., Martinez, G. A., Scoma, A., Fraraccio, S., Kerckhof, F.-M., Boon, N., Reis, M. A. M., Fava, F. & Bertin, L. 2017 Effect of operational parameters in the continuous anaerobic fermentation of cheese whey on titers, yields, productivities, and microbial community structures. *ACS Sustainable Chemistry & Engineering* **5** (2), 1400–1407. https://doi.org/10.1021/acssuschemeng.6b01901.

EAWAG 2019 Forum Chriesbach. EAWAG. 2019. Available from: https://www.eawag.ch/en/aboutus/sustainability/sustainable-building/forum-chriesbach/.

Ellen MacArthur Foundation 2017 *Urban Biocycles.* Available from: https://www.ellenmacarthurfoundation.org/publications/urban-biocycles.

EPEM S.A. – Environmental Planning, Engineering & Management 2011 *Database of Waste Management Technologies*. Available from: http://www.epem.gr/waste-c-control/database/html/MBT-07.htm.

Eregowda, T., Matanhike, L., Rene, E. R. & Lens, P. N. L. 2018 Performance of a biotrickling filter for the anaerobic utilization of gas-phase methanol coupled to thiosulphate reduction and resource recovery through volatile fatty acids production. *Bioresource Technology* **263**, 591–600. https://doi.org/10.1016/j.biortech.2018.04.095.

Etter, B., Udert, K. M. & Hug, A. 2013 Total nutrient recovery from urine – operation of a pilot-scale nitrification reactor. In: *WEF/IWA International Conference on Nutrient Removal and Recovery*, Vancouver, Canada. Available from: https://www.researchgate.net/publication/265966790_Total_Nutrient_Recovery_from_Urine_-_Operation_of_a_Pilot-Scale_Nitrification_Reactor.

EUBIA 2019 *Bioethanol – European Biomass Industry Association*. Available from: http://www.eubia.org/cms/wiki-biomass/biofuels/bioethanol/.

European Commission 1991 *Official Journal of the European Communities*, **34**(May). Available from: https://eur-lex.europa.eu/legal-content/EN/TXT/?uri=OJ:L:1991:135:TOC.

European Commission 2018a *A Sustainable Bioeconomy for Europe: Strengthening the Connection Between Economy, Society and the Environment*. Publications Office of the European Union, Luxembourg.

European Commission 2018b Water Reuse. *Environment* (blog). May 28, 2018. Available from: http://ec.europa.eu/environment/water/reuse.htm.

European Commission 2019 Municipal Waste Statistics. *Eurostat Statistics Explained* (blog). January 23, 2019. Available from: https://ec.europa.eu/eurostat/statistics-explained/index.php/Municipal_waste_statistics.

European Commission – Eurostat 2019 Monitoring Framework. Available from: https://ec.europa.eu/eurostat/web/circular-economy/indicators/monitoring-framework.

Fermoso, F. G., van Hullebusch, E., Collins, G., Roussel, J., Mucha, A. P. & Esposito, G. (eds) 2019 *Trace Elements in Anaerobic Biotechnologies*. IWA Publishing, London, UK. https://doi.org/10.2166/9781789060225.

Fernandez-Cañero, R., Perez Urrestarazu, L. & Franco Salas, A. 2012 Assessment of the cooling potential of an indoor living wall using different substrates in a warm climate. *Indoor and Built Environment* **21** (5), 642–650.

Fertilizers Europe 2017 *Industry Facts and Figures 2018*. Fertilizers Europe, asbl, Brussels, Belgium. Available from: http://fertilizerseurope.com/.

Finger, D., Wüest, A. & Bossard, P. 2013 Effects of oligotrophication on primary production in peri-alpine lakes: modeling primary production in lakes. *Water Resources Research* **49** (8), 4700–4710. https://doi.org/10.1002/wrcr.20355.

Garcia-Aguirre, J., Aymerich, E., González-Mtnez de Goñi, J. & Esteban-Gutiérrez, M. 2017 Selective VFA production potential from organic waste streams: assessing temperature and pH influence. *Bioresource Technology* **244**, 1081–1088. https://doi.org/10.1016/j.biortech.2017.07.187.

Gattringer, H., Claret, A., Radtke, M., Kisser, J., Zraunig, A., Rodriguez-Roda, I. & Buttiglieri, G. 2016 Novel vertical ecosystem for sustainable water treatment and reuse in tourist resorts. *International Journal of Sustainable Development and Planning* **11** (3), 263–274. https://doi.org/10.2495/SDP-V11-N3-263-274.

Goethem, C. V., Mertens, M., Cirujano, F. G., Seo, J. W., De Vos, D. & Vankelecom, I. F. J. 2018 Improved MOF nanoparticle recovery and purification using crosslinked PVDF membranes. *Chemical Communications* **54** (53), 7370–7373. https://doi.org/10.1039/C8CC04326D.

Green City Solutions 2019 *Green City Solutions*. Available from: https://greencitysolutions.de/en/.

Guardian 2019 From No Recycling to Zero Waste: How Ljubljana Rethought Its Rubbish. *The Guardian*, May 23, 2019, sec. Cities. Available from: https://www.theguardian.com/cities/2019/may/23/zero-recycling-to-zero-waste-how-ljubljana-rethought-its-rubbish.

Guenther, K., Heinke, V., Thiele, B., Kleist, E., Prast, H. & Raecker, T. 2002 Endocrine disrupting nonylphenols are ubiquitous in food. *Environmental Science & Technology* **36** (8), 1676–1680.

Hennebel, T., Boon, N., Maes, S. & Lenz, M. 2015 Biotechnologies for critical raw material recovery from primary and secondary sources: R&D priorities and future perspectives. *New Biotechnology* **32** (1), 121–127. https://doi.org/10.1016/j.nbt.2013.08.004.

Hoelzle, R. D., Virdis, B. & Batstone, D. J. 2014 Regulation mechanisms in mixed and pure culture microbial fermentation. *Biotechnology and Bioengineering* **111** (11), 2139–2154. https://doi.org/10.1002/bit.25321.

Houston, J., Casazza, E., Briguglio, M. & Spiteri, J. 2018 *Enablers and Barriers to A Circular Economy*. R2PI H2020 Project Deliverable. Stakeholder Views Report. Available from: http://www.r2piproject.eu/wp-content/uploads/2018/08/R2pi-stakeholders-report-sept-2018.pdf.

Houzelot, V., Laubie, B., Pontvianne, S. & Simonnot, M.-O. 2017 Effect of up-scaling on the quality of ashes obtained from hyperaccumulator biomass to recover Ni by agromining. *Chemical Engineering Research and Design* **120**, 26–33. https://doi.org/10.1016/j.cherd.2017.02.002.

Houzelot, V., Ranc, B., Laubie, B. & Simonnot, M.-O. 2018 Agromining of hyperaccumulator biomass: study of leaching kinetics of extraction of nickel, magnesium, potassium, phosphorus, iron, and manganese from *Alyssum murale* ashes by sulfuric acid. *Chemical Engineering Research and Design* **129**, 1–11. https://doi.org/10.1016/j.cherd.2017.10.030.

IBA Hamburg GmbH 2013 *BIQ – Smart Material Houses*. Available from: https://www.iba-hamburg.de/projekte/bauausstellung-in-der-bauausstellung/smart-material-houses/biq/projekt/biq.html.

Istenič, D., Božič, G., Arias, C. A. & Griessler Bulc, T. 2017 Growth dynamic of three different white willow clones used in a zero-discharge wastewater treatment system in the sub-Mediterranean region – an early evaluation. *Desalination and Water Treatment* **91**, 260–267. https://doi.org/10.5004/dwt.2017.21186.

Istenič, D., Arias, C. A., Pavliha, G. & Griessler Bulc, T. 2018 Evapotranspiration and biomass production in a willow system under sub-Mediterranean climate. In: *IWA Conference Proceedings*, Valencia, Spain.

Jankowska, E., Chwialkowska, J., Stodolny, M. & Oleskowicz-Popiel, P. 2017 Volatile fatty acids production during mixed culture fermentation – the impact of substrate complexity and pH. *Chemical Engineering Journal* **326**, 901–910. https://doi.org/10.1016/j.cej.2017.06.021.

Kabir, E. R., Sharfin Rahman, M. & Rahman, I. 2015 A review on endocrine disruptors and their possible impacts on human health. *Environmental Toxicology and Pharmacology* **40** (1), 241–258. https://doi.org/10.1016/j.etap.2015.06.009.

Katsou, E., Nika, C.-E., Buehler, D., Maric, B., Megyesi, B., Mino, E., Almenar, J. B. & Bas, B. 2019 Transformation tools enabling the implementation of nature-based solutions for creating a resourceful circular city. *Blue-Green Systems*. https://doi.org/10.2166/bgs.2020.929.

Khan, M. I., Shin, J. H. & Kim, J. D. 2018 The promising future of microalgae: current status, challenges, and optimization of a sustainable and renewable industry for biofuels, feed, and other products. *Microbial Cell Factories* **17** (March). https://doi.org/10.1186/s12934-018-0879-x.

Kibler, K. M., Reinhart, D., Hawkins, C., Mohaghegh Motlagh, A. & Wright, J. 2018 Food waste and the food-energy-water nexus: a review of food waste management alternatives. *Waste Management* **74**, 52–62. https://doi.org/10.1016/j.wasman.2018.01.014.

Langergraber, G., Pucher, B., Simperler, L., Kisser, J., Katsou, E., Buehler, D., Garcia Mateo, M. C. & Atanasova, N. 2019 Implementing nature-based solutions for creating a resourceful circular city. *Blue-Green Systems* 2 (1), 173–185. https://doi.org/10.2166/bgs.2020.933.

Larsen, T. A. 2011 Redesigning wastewater infrastructure to improve resource efficiency. *Water Science and Technology* **63** (11), 2535–2541. https://doi.org/10.2166/wst.2011.502.

Larsen, T. A., Gebauer, H., Gründl, H., Künzle, R., Lüthi, C., Messmer, U., Morgenroth, E., Niwagaba, C. B. & Ranner, B. 2015 Blue diversion: a new approach to sanitation in informal settlements. *Journal of Water Sanitation and Hygiene for Development* **5** (1), 64–71. https://doi.org/10.2166/washdev.2014.115.

Lettinga, G., Van Velsen, A. F. M., Hobma, S. & De Zeeuw, W. 1981 Feasibility of anaerobic digestion for the direct treatment of, and the energy recovery from urban wastes. *Studies in Environmental Science* **9**, 97–109. https://doi.org/10.1016/S0166-1116(08)71356-6.

Li, Y. M., Chaney, R., Brewer, E., Roseberg, R., Angle, J. S., Baker, A., Reeves, R. & Nelkin, J. 2003 Development of a technology for commercial phytoextraction of nickel: economic and technical considerations. *Plant and Soil* **249** (1), 107–115.

Li, Y., Horsman, M., Wu, N., Lan, C. Q. & Dubois-Calero, N. 2008 Biofuels from microalgae. *Biotechnology Progress* **24** (4), 815–820. https://doi.org/10.1021/bp070371 k.

Logan, M. & Visvanathan, C. 2019 Management strategies for anaerobic digestate of organic fraction of municipal solid waste: current status and future prospects. *Waste Management & Research* **37** (1 suppl.), 27–39. https://doi.org/10.1177/0734242X18816793.

Madigan, M. T. & Jung, D. O. 2009 An overview of purple bacteria: Systematics, physiology, and habitats. In: *The Purple Phototrophic Bacteria*. Advances in Photosynthesis and Respiration (Neil Hunter, C., Daldal, F., Thurnauer, M. C. & Thomas Beatty, J., eds). Springer, Dordrecht, The Netherlands, pp. 1–15. https://doi.org/10.1007/978-1-4020-8815-5_1.

Malinska, K. & Zabochnicka-Swiatek, M. 2010 Biosystems for air protection. In: *Air Pollution* (Villanyi, V., ed.). InTech. https://doi.org/10.5772/10048.

Mansouri, S. S., Udugama, I. A., Cignitti, S., Mitic, A., Flores-Alsina, X. & Gernaey, K. V. 2017 Resource recovery from bio-based production processes: a future necessity? *Current Opinion in Chemical Engineering* **18**, 1–9. https://doi.org/10.1016/j.coche.2017.06.002.

Martí-Herrero, J., Soria-Castellón, G., Diaz-de-Basurto, A., Alvarez, R. & Chemisana, D. 2019 Biogas from a full scale digester operated in psychrophilic conditions and fed only with fruit and vegetable waste. *Renewable Energy* **133** (April), 676–684. https://doi.org/10.1016/j.renene.2018.10.030.

Mauky, E., Weinrich, S., Jacobi, H.-F., Nägele, H.-J., Liebetrau, J. & Nelles, M. 2017 Demand-driven biogas production by flexible feeding in full-scale – process stability and flexibility potentials. *Anaerobe* **46**, 86–95. https://doi.org/10.1016/j.anaerobe.2017.03.010.

Miklosi, A. 2013 *Chlorella Oxygen Pavilion Concept and the Symbiosis in Design*. Göteborgs Universitetsbibliotek, Gothenburg, Sweden. Available from: https://gupea.ub.gu.se/handle/2077/32306.

Mo, W. Y., Man, Y. B. & Wong, M. H. 2018 Use of food waste, fish waste and food processing waste for China's aquaculture industry: needs and challenge. *The Science of the Total Environment* **613–614**, 635–643. https://doi.org/10.1016/j.scitotenv.2017.08.321.

Morais, M. G. & Costa, J. A. V. 2007 Carbon dioxide fixation by *Chlorella kessleri, C. vulgaris, Scenedesmus obliquus* and *Spirulina* sp. cultivated in flasks and vertical tubular photobioreactors. *Biotechnology Letters* **29**, 1349–1352.

Murovec, B., Kolbl, S. & Stres, B. 2015 Methane yield database: online infrastructure and bioresource for methane yield data and related metadata. *Bioresource Technology* **189**, 217–223. https://doi.org/10.1016/j.biortech.2015.04.021.

Nehls, T., Jiang, Y., Dennehy, C., Zhan, X. & Beesley, L. 2015 From waste to value: urban agriculture enables cycling of resources in cities. In: *Urban Agriculture Europe* (Lohrberg, F., Licka, L., Scazzosi, L. & Timpe, A., eds). JOVIS, Berlin, Germany, pp. 170–173.

Ni, T. W., Staicu, L. C., Nemeth, R. S., Schwartz, C. L., Crawford, D., Seligman, J. D., Hunter, W. J., Pilon-Smits, E. A. H. & Ackerson, C. J. 2015 Progress toward clonable inorganic nanoparticles. *Nanoscale* **7** (41), 17320–17327. https://doi.org/10.1039/c5nr04097c.

Nordin, A. C. & Vinnerås, B. 2015 Sanitising black water by auto-thermal aerobic digestion (ATAD) combined with ammonia treatment. *Water Science and Technology* **72** (12), 2112–2121. https://doi.org/10.2166/wst.2015.432.

Nordin, A., Göttert, D. & Vinnerås, B. 2018 Decentralised black water treatment by combined auto-thermal aerobic digestion and ammonia – a pilot study optimising treatment capacity. *Journal of Environmental Management* **207**, 313–318. https://doi.org/10.1016/j.jenvman.2017.10.064.

O'Dwyer, E., Wang, H., Wang, A., Shah, N. & Guo, M. 2018 Optimisation of wastewater treatment and recovery solutions in industrial parks. *Computer Aided Chemical Engineering* **43**, 1407–1412. https://doi.org/10.1016/B978-0-444-64235-6.50246-1.

Osmani, M. & Bani, A. 2017 Heavy metals concentration of dumping site soils and their accumulation in Alyssum murale growing in selected dumping sites in Albania. *Thalassia Salentina* **39**. https://doi.org/10.1285/i15910725v39p83.

Osmani, M., Bani, A. & Hoxha, B. 2018a The phytomining of nickel from industrial polluted site of Elbasan-Albania. *European Academic Research* **V** (10). Available from: https://www.researchgate.net/publication/323457449_The_Phytomining_of_nickel_from_industrial_polluted_site_of_Elbasan-Albania.

Osmani, M., Bani, A., Gjoka, F., Pavlova, D., Naqellari, P., Shahu, E., Duka, I. & Echevarria, G. 2018b The natural plant colonization of ultramafic post-mining area of Përrenjas, Albania. *Periodico Di Mineralogia* **87** (2), 135–146. https://doi.org/10.2451/2018PM729.

Pandey, V. C. & Souza-Alonso, P. 2019 Market opportunities: in sustainable phytoremediation. In: *Phytomanagement of Polluted Sites* (Pandey, V. C. & Bauddh, K., eds). Elsevier, pp. 51–82. https://doi.org/10.1016/B978-0-12-813912-7.00002-8.

Pascual, A., De la Varga, D., Soto, M., Van Oirschot, D., Kilian, R. M., Álvarez, J. A., Carvalho, P., Brix, H. & Arias, C. A. 2018 Aerated constructed wetlands for treatment of municipal and food industry wastewater. In: *Constructed Wetlands for Industrial Wastewater Treatment* (Stefanakis, A. I., ed). John Wiley & Sons, Ltd, pp. 65–93. https://doi.org/10.1002/9781119268376.ch3.

Passos, F., Astals, S. & Ferrer, I. 2014 Anaerobic digestion of microalgal biomass after ultrasound pretreatment. *Waste Management* **34** (11), 2098–2103. https://doi.org/10.1016/j.wasman.2014.06.004.

Pettit, T., Irga, P. J. & Torpy, F. R. 2018 Towards practical indoor air phytoremediation: a review. *Chemosphere* **208**, 960–974. https://doi.org/10.1016/j.chemosphere.2018.06.048.

Plaza, M., Herrero, M., Cifuentes, A. & Ibáñez, E. 2009 Innovative natural functional ingredients from microalgae. *Journal of Agricultural and Food Chemistry* **57** (16), 7159–7170. https://doi.org/10.1021/jf901070 g.

Puyol, D., Batstone, D. J., Hülsen, T., Astals, S., Peces, M. & Krömer, J. O. 2017 Resource recovery from wastewater by biological technologies: opportunities, challenges, and prospects. *Frontiers in Microbiology* **7**. https://doi.org/10.3389/fmicb.2016.02106.

Raganati, F., Procentese, A., Olivieri, G., Russo, M. E., Salatino, P. & Marzocchella, A. 2014 MFA of clostridium acetobutylicum pathway: the role of glucose and xylose on the acid formation/uptake. *Chemical Engineering Transactions* **38**, 337–342.

Randall, D. G., Krähenbühl, M., Köpping, I., Larsen, T. A. & Udert, K. M. 2016 A novel approach for stabilizing fresh urine by calcium hydroxide addition. *Water Research* **95**, 361–369. https://doi.org/10.1016/j.watres.2016.03.007.

Razzak, S. A., Hossain, M. M., Lucky, R. A., Bassi, A. S. & de Lasa, H. 2013 Integrated CO_2 capture, wastewater treatment and biofuel production by microalgae culturing – a review. *Renewable and Sustainable Energy Reviews* **27**, 622–653. https://doi.org/10.1016/j.rser.2013.05.063.

Rijksdienst voor Ondernemend Nederland 2019 Greendeals. Available from: https://www.greendeals.nl/english.

Rodenas Motos, P., Ter Heijne, A., van der Weijden, R., Saakes, M., Buisman, C. H. & Sleutels, T. H. 2015 High rate copper and energy recovery in microbial fuel cells. *Frontiers in Microbiology* **6**, 527. https://doi.org/10.3389/fmicb.2015.00527.

Rodenas Motos, P., Molina, G., Ter Heijne, A., Sleutels, T., Saakes, M. & Buisman, C. 2017 Prototype of a scaled-up microbial fuel cell for copper recovery. *Journal of Chemical Technology and Biotechnology* **92** (11), 2817–2824. https://doi.org/10.1002/jctb.5353.

Saadiah, H., Aini, N., Rahman, A., Kalsom, U., Baharuddin, S. & Ari, A. B. 2017 Feasibility of using kitchen waste as future substrate for bioethanol production: a review. *Renewable and Sustainable Energy Reviews* **74**, 671–686. https://doi.org/10.1016/j.rser.2017.02.071.

Senecal, J. & Vinnerås, B. 2017 Urea stabilisation and concentration for urine-diverting dry toilets: urine dehydration in ash. *Science of the Total Environment* **586**, 650–657. https://doi.org/10.1016/j.scitotenv.2017.02.038.

Sharkey, T. D., Wiberley, A. E. & Donohue, A. R. 2008 Isoprene emission from plants: why and how. *Annals of Botany* **101** (1), 5–18. https://doi.org/10.1093/aob/mcm240.

Sigma-Aldrich 2018 *Ammonium Nickel(II) Sulfate Hexahydrate*. Available from: https://www.sigmaaldrich.com.

Simha, P., Senecal, J., Nordin, A., Lalander, C. & Vinnerås, B. 2018 Alkaline dehydration of anion-exchanged human urine: volume reduction, nutrient recovery and process optimisation. *Water Research* **142**, 325–336. https://doi.org/10.1016/j.watres.2018.06.001.

Simonnot, M.-O., Vaughan, J. & Laubie, B. 2018 Processing of bio-ore to products BT. In: *Agromining: Farming for Metals: Extracting Unconventional Resources Using Plants* (Van der Ent, A., Echevarria, G., Baker, A. J. M. & Morel, J. L., eds). Springer International Publishing, Cham, Switzerland, pp. 39–51. https://doi.org/10.1007/978-3-319-61899-9_3.

Singh, S., Rinta-Kanto, J. M., Kettunen, R., Lens, P., Collins, G., Kokko, M. & Rintala, J. 2019 Acetotrophic activity facilitates methanogenesis from LCFA at low temperatures: screening from mesophilic inocula. *Archaea* 1–16. https://doi.org/10.1155/2019/1751783.

Slavov, A. K. 2017 General characteristics and treatment possibilities of dairy wastewater – a review. *Food Technology and Biotechnology* **55** (1), 14–28. https://doi.org/10.17113/ft b.55.01.17.4520.

Song, X., Luo, W., Hai, F. I., Price, W. E., Guo, W., Ngo, H. H. & Nghiem, L. D. 2018 Resource recovery from wastewater by anaerobic membrane bioreactors: opportunities and challenges. *Bioresource Technology* **270**, 669–677. https://doi.org/10.1016/j.biortech.2018.09.001.

Stichting Toegepast Onderzoek Waterbeheer - Rijkskantoor Rijnstraat 8. Saniwijzer 2017 Available from: https://www.saniwijzer.nl/projecten/rijkskantoor-rijnstraat-8/detail=94 (accessed 3 April 2019).

Strazzera, G., Battista, F., Herrero Garcia, N., Frison, N. & Bolzonella, D. 2018 Volatile fatty acids production from food wastes for biorefinery platforms: a review. *Journal of Environmental Management* **226**, 278–288. https://doi.org/10.1016/j.jenvman.2018.08.039.

Su, Y. M. & Lin, C. H. 2015 Removal of indoor carbon dioxide and formaldehyde using green walls by bird nest fern. *Horticulture Journal* **84** (1), 69–76.

Sulonen, M. L. K., Kokko, M. E., Lakaniemi, A.-M. & Puhakka, J. A. 2018 Simultaneous removal of tetrathionate and copper from simulated acidic mining water in bioelectrochemical and electrochemical systems. *Hydrometallurgy* **176**, 129–138. https://doi.org/10.1016/j.hydromet.2018.01.023.

Tamis, J., Mulders, M., Dijkmans, H., Rozendal, R., van Loosdrecht, M. C. M. & Kleerebezem, R. 2018 Pilot-scale polyhydroxyalkanoate production from paper mill wastewater: process characteristics and identification of bottlenecks for full-scale implementation. *Journal of Environmental Engineering* **144** (10), 04018107. https://doi.org/10.1061/(ASCE)EE.1943-7870.0001444.

Tang, J., Zhang, C., Shi, X., Sun, J. & Cunningham, J. A. 2019 Municipal wastewater treatment plants coupled with electrochemical, biological and bio-electrochemical technologies: opportunities and challenge toward energy self-sufficiency. *Journal of Environmental Management* **234**, 396–403. https://doi.org/10.1016/j.jenvman.2018.12.097.

Techmetall 2019 *grünwand Klimafassade*. Available from: https://gruenwand.com/.

Trulli, E., Ferronato, N., Torretta, V., Piscitelli, M., Masi, S. & Mancini, I. 2018 Sustainable mechanical biological treatment of solid waste in urbanized areas with low recycling rates. *Waste Management* **71**, 556–564. https://doi.org/10.1016/j.wasman.2017.10.018.

Udert, K. M. & Wächter, M. 2012 Complete nutrient recovery from source-separated urine by nitrification and distillation. *Water Research* **46** (2), 453–464. https://doi.org/10.1016/j.watres.2011.11.020.

Uwizeye, M. L., Venditti, S. & Hansen, J. 2017 Selection criteria of Microthrix parvicella as lipid accumulator in activated sludge treatment plants. In: *IWA YWP Conference BeNeLux 2017*, Luxembourg.

Van der Roest, H., Van Loosdrecht, M., Langkamp, E. J. & Uijterlinde, C. 2015 Recovery and reuse of alginate from granular Nereda sludge. *Water* **21**, 48.

Vlaeminck, S. E., Terada, A., Smets, B. F., Van der Linden, D., Boon, N., Verstraete, W. & Carballa, M. 2009 Nitrogen removal from digested black water by one-stage partial nitritation and anammox. *Environmental Science & Technology* **43** (13), 5035–5041. https://doi.org/10.1021/es803284y.

VUNA GmbH 2019 Available from: http://vuna.ch/.

Wang, H., Xu, J. & Sheng, L. 2019 Study on the comprehensive utilization of city kitchen waste as a resource in China. *Energy* **173**. https://doi.org/10.1016/j.energy.2019.02.081.

Waterschap Rijn en IJssel 2018 *Kaumera* (blog). Available from: https://kaumera.com/english/kaumera/.

Wei, X., Lyu, S., Yu, Y., Wang, Z., Liu, H., Pan, D. & Chen, J. 2017 Phylloremediation of air pollutants: exploiting the potential of plant leaves and leaf-associated microbes. *Frontiers in Plant Science* **8**. https://doi.org/10 3389/fpls.2017.01318.

Weißbach, M., Thiel, P., Drewes, J. E. & Koch, K. 2018 Nitrogen removal and intentional nitrous oxide production from reject water in a coupled nitritation/nitrous denitritation system under real feed-stream conditions. *Bioresource Technology* **255**, 58–66. https://doi.org/10.1016/j.biortech.2018.01.080.

Weithmann, N., Möller, J. N., Löder, M. G. J., Piehl, S., Laforsch, C. & Freitag, R. 2018 Organic fertilizer as a vehicle for the entry of microplastic into the environment. *Science Advances* **4** (4), eaap8060. https://doi.org/10.1126/sciadv.aap8060.

Wit, J. B. d., de Graaf, R., Elzinga, N., Debucquoy, W., 't Lam, G. P., Rodenhuis, J., Stutterheim, E. & Piekema, L. 2018 *Evaluatie Nieuwe Sanitation Noorderhoek/Waterschoon 2 (Evaluation New Sanitation Noorderhoek/Waterschoon 2)*. STOWA, Amersfoort, The Netherlands.

Wolverton, B. C. & McDonald, C. R. 1983 *Microbial Filters and Higher Plants for Treating Hazardous and Toxic Chemicals*. Available from: http://airex.tksc.jaxa.jp/pl/dr/19930073014/en.

Wolverton, B. C. & McDonald-McCaleb, R. C. 1986 *Biotransformation of Priority Pollutants Using Biofilms and Vascular Plants*. EPA/600/J-86/310. U.S. Environmental Protection Agency, Washington, DC, USA. Available from: http://ntrs.nasa.gov/search.jsp?N=0&Ntk=all&Ntx=mode matchall&Ntt=20050236283.

Wolverton, B. C. & Wolverton, J. D. 1993 Apparatus for purifying waste water and air in an indoor environment, issued 1993. United States Patent US5269094A. Available from: http://www.google.com/patents?hl=en&lr=&vid=USPAT5269094&id=KIwlAAAAEBAJ&oi=fnd&dq=Apparatus+for+purifying+waste+water+and+air+in+an+indoor+environment&printsec=abstract.

Wood, R. A., Burchett, M. D., Alquezar, R., Orwell, R. L., Tarran, J. & Torpy, F. 2006 The potted-plant microcosm substantially reduces indoor air VOC pollution: I. Office field-study. *Water, Air, and Soil Pollution* **175** (1–4), 163–180. https://doi.org/10.1007/s11270-006-9124-z.

World Bank 2017 *Urban Development Overview.* World Bank, Washington, DC, USA. Available from: http://www.worldbank.org/en/topic/urbandevelopment/overview.

World Health Organization 2011 *WHO Guidelines for Drinking-Water Quality.* World Health Organization, Geneva, Switzerland. Available from: https://books.google.com/books?hl=de&lr=&id=SJ76COTm-nQC&oi=fnd&pg=PR15&dq=WHO+Guidelines+for+Drinking-water+Quality:+Recommendations&ots=V7sXreT8Yg&sig=qjy6jMaeBxK-nkvotmfipoZudZg.

Wuang, S. C., Khin, M. C., Chua, P. Q. D. & Luo, Y. D. 2016 Use of *Spirulina* biomass produced from treatment of aquaculture wastewater as agricultural fertilizers. *Algal Research* **15**, 59–64. https://doi.org/10.1016/j.algal.2016.02.009.

Xu, Z., Wang, L. & Hou, H. 2011 Formaldehyde removal by potted plant-soil systems. *Journal of Hazardous Materials* **192** (1), 314–318. https://doi.org/10.1016/j.jhazmat.2011.05.020.

Zacharof, M.-P. & Lovitt, R. W. 2014 Recovery of volatile fatty acids (VFA) from complex waste effluents using membranes. *Water Science and Technology* **69** (3), 495–503. https://doi.org/10.2166/wst.2013.717.

Zeeman, G. 2012 *New Sanitation: Bridging Cities and Agriculture.* Wageningen University, Wageningen, The Netherlands.

Zeeman, G., Kujawa, K., de Mes, T. Z. D., de Graaff, M. S., Abu-Ghunmi, L. N. A. H., Mels, A. R., Meulman, B., Temmink, H., Buisman, C., van Lier, J. & Lettinga, G. 2008 Anaerobic treatment as a core technology for energy, nutrients and water from source-separated domestic waste(water). *Water Science and Technology* **57** (8), 1207–1212.

Zhang, X., Laubie, B., Houzelot, V., Plasari, E., Echevarria, G. & Simonnot, M.-O. 2016 Increasing purity of ammonium nickel sulfate hexahydrate and production sustainability in a nickel phytomining process. *Chemical Engineering Research and Design* **106**, 26–32. https://doi.org/10.1016/j.cherd.2015.12.009.

Zraunig, A., Estelrich, M., Gattringer, H., Kisser, J., Langergraber, G., Radtke, M., Rodriguez-Roda, I. & Buttiglieri, G. 2019 Long term decentralized greywater treatment for water reuse purposes in a tourist facility by vertical ecosystem. *Ecological Engineering* **138**, 138–147. https://doi.org/10.1016/j.ecoleng.2019.07.003.

First received 30 June 2019; accepted in revised form 15 October 2019. Available online 3 February 2020

doi: 10.2166/bgs.2019.931

Urban agriculture as a keystone contribution towards securing sustainable and healthy development for cities in the future

S. L. G. Skar[a,*], R. Pineda-Martos[b], A. Timpe[c], B. Pölling[d], K. Bohn[e], M. Külvik[f], C. Delgado[g], C. M.G. Pedras[h], T. A. Paço[i], M. Ćujić[j], N. Tzortzakis[k], A. Chrysargyris[k], A. Peticila[l], G. Alencikiene[m], H. Monsees[n] and R. Junge IWA [o]

[a] Division of Food Production and Society, Dept. of Horticulture, Norwegian Institute of Bioeconomic Research (NIBIO), P.O. BOX 115, 1431 Aas/Reddalsveien 215, 4886 Grimstad, Norway

[b] School of Agricultural Engineering, University of Sevilla, Sevilla, Spain

[c] RWTH Aachen University (Rheinisch-Westfälische Technische Hochschule Aachen), Germany

[d] South-Westphalia University of Applied Sciences, Soest, Germany

[e] School of Architecture and Design, University of Brighton, UK

[f] Institute of Agricultural and Environmental Sciences, Estonian University of Life Sciences, Friedrich Reinhold Kreutzwaldi 1, Tartu 51006, Estonia

[g] Faculdade de Ciências Sociais e Humanas – Interdisciplinary Center of Social Sciences – CICS.NOVA, Universidade Nova de Lisboa, Av. de Berna, 26-C, 1069-061 Lisboa, Portugal

[h] Faculdade de Ciências e Tecnologia, University of Algarve, Faro, Portugal

[i] LEAF, Linking Landscape, Environment, Agriculture and Food, Instituto Superior de Agronomia, Universidade de Lisboa, Lisboa, Portugal

[j] Vinca Institute of Nuclear Sciences, University of Belgrade, Mice Petrovica Alasa, Belgrade, Serbia

[k] Department of Agricultural Sciences, Biotechnology and Food Science, Cyprus University of Technology, Anexartisias 33, P.O. BOX 50329, 3603, Limassol, Cyprus

[l] USAMV Bucharest, Marasti Blv 59, Bucharest, Romania

[m] Kaunas University of Technology, Food Institute, K. Donelaičio St. 73, 44249 Kaunas, Lithuania

[n] Dpt. Ecophysiology and Aquaculture, Leibniz Institute of Freshwater Ecology and Inland Fisheries, Müggelseedamm 310, 12587 Berlin, Germany

[o] IUNR Institute of Natural Resource Sciences, ZHAW Zurich University of Applied Sciences, P.O. Box, CH-8820 Wädenswil, Switzerland

*Corresponding author. E-mail: siv.skar@nibio.no

Abstract

Research and practice during the last 20 years has shown that urban agriculture can contribute to minimising the effects of climate change by, at the same time, improving quality of life in urban areas. In order to do so most effectively, land use and spatial planning are crucial so as to obtain and maintain a supportive green infrastructure and to secure citizens' healthy living conditions. As people today trend more towards living in green and sustainable city centres that can offer fresh and locally produced food, cities become again places for growing food. The scope of urban agriculture thereby is to establish food production sites within the city's sphere; for example, through building-integrated agriculture including concepts such as aquaponics, indoor agriculture, vertical farming, rooftop production, edible walls, as well as through urban farms, edible landscapes, school gardens and community gardens. Embedded in changing urban food systems, the contribution of urban agriculture to creating sustainable and climate-friendly cities is pivotal as it has the capacity to integrate other resource streams such as water, waste and energy. This article describes some of the current aspects of the circular city debate where urban agriculture is pushing forward the development of material and resource cycling in cities.

Key words: agriculture, circular city, ecosystem services, infrastructure, recirculation, urban farming

INTRODUCTION

Humans today face a plethora of environmental challenges. Some of them are aggravated by concentration of the populations in the cities, all of which constitute a complex set of future challenges; transport of people and goods (i.e. excessive carbon emissions), need for substantial rainfall absorption, high levels of air pollution, urban heat island effect, drinking water supply, waste management, lack of biodiversity. All these challenges result in illness and stress syndromes in the population. Urban agriculture has the potential to contribute towards minimising several of these adverse effects and thus improve the liveability of cities. This review aims to address and clarify some of these aspects, in order to facilitate the implementation of urban agriculture within the context of nature-based solutions (NBS) in circular cities of the future.

Currently, the majority of the world population growth is in the cities, especially in developing countries. Urban areas worldwide are expected to absorb all the population growth expected over the next four decades and continue to draw in the rural population (United Nations 2018). While cities today cover about 2–3% of all land area, they consume approximately 75% of the world's energy and generate 80% of the CO_2 emissions (UN 2018). The cities also utilise large quatities of water, create an enormous quantity of waste and pollute the air. Climate changes are predicted to cause more environmental stressors in the future, while we need to intensify food production (Junge & Graber 2014). The required transition will need increased flexibility of the urban environment, more sustainable use and re-use of natural resources as well as the adaptation of infrastructure systems (Herrera-Gomez *et al.* 2017). All this requires future city development to be smart and to integrate innovative solutions. One key to a more sustainable and healthy city development in the future might be in a relocalisation of the food system and a narrowing of the cities' foodsheds. This idea perfectly coincides with the idea of a circular city, where organic disposals are reused as resources to produce new agricultural products. The contemporary linear understanding of a city, where most independent entities consume, metabolise and dispose of resources, urgently needs – not only but especially in the field of food – a more systematic perspective to solve existing challenges.

In contrast to other infrastructures like water and electricity, the food production, and provisioning system did not get much attention in city planning and are still a neglected field (Pothukuchi & Kaufman 1999). It needs to be considered that in the process of building the modern city historical ties and links to the localised food system have been disrupted. During expansion, cities lost large areas of their surrounding fertile farmland and have mostly benefited from access to a globalised food system. The consequences of the globalisation of the food system can be seen in an abundance of food but also in the creation of a not sustainable industrialised system that overproduces, pollutes natural resources, declines biodiversity and stimulate obesity and malnourishment (Kennedy *et al.* 2004; IPBES 2019). It has also favoured massive path dependencies (Moragues-Faus *et al.* 2017) that can have a significant impact on food security and that needs to be identified and overcome to shape localised, circular economy-based food systems for the circular city.

doi: 10.2166/bgs.2019.931

Urban environment conditions, such as air quality, solar radiation and climate are inherently different from rural environments, and these differences may have an impact on crop growth (Eriksen-Hamel & Danso 2010). One of the risks for urban farming that stems from air pollution is decreased irradiance, caused by solar dimming, which is caused by increased reflectance of radiation away from the ground, due to air pollutants and aerosols over urban areas. Polluted urban areas can receive 8% less solar radiation than rural areas (Eriksen-Hamel & Danso 2010).

This article summarises major aspects related to urban agriculture in order to implement circular agriculture-based schemes in urban settings. What resources exist in an urban biosphere to bring into the context of urban agriculture? The main purpose of urban farming is to produce food within a city, but we also want to pay closer attention to other resources available from urban farming systems, which are usually considered to be waste.

URBAN AGRICULTURE AND CIRCULAR CITIES

Urban food systems: urban agriculture and urban farming

As an increasingly popular phenomenon in different domains during the last twenty years, Urban Agriculture (Table 1) is discussed in science, policy-making, media and society and its definition is context-dependent (Delgado 2018). Because of the rapid development of the field, several interpretations of the term 'urban agriculture' exist, capturing nuances within different contexts. Amongst those, two definitions stand out: one from the seminal publication of the United Nations Development Program (UNDP) (Smit 1996); and the other from Mougeot (2001) which provides an extension of the former stressing that it is 'its integration into the urban economic and ecological system' that distinguishes urban from rural agriculture rather than its urban location only.

Smit's and Mougeot's definitions are nowadays the most commonly used ones and are valued for their simplicity, openness and implicit inclusion of a circular-city approach.

Urban agriculture spans all actors, communities, activities, places, and economies which focus on primary production in a spatial context categorised as 'urban' (Vejre *et al.* 2015), it can be structured into two sub-groups; urban food gardening and urban farming (Table 1). The common denominator of both is the bio-based output of products, which are harvested and consumed while other effects of urban agriculture on environment and society can be classified as by-products, externalities or co-benefits. These benefits can be classified into four dimensions: food security, economic, social and environmental (McEldowney 2017), and include contributing to employment, improved education and health, to social inclusion through integrating those at risk of social exclusion. *The business-as-usual farming* operation, as well as non-urban adapted farming, also exist in urban areas (Deelstra & Girardet 2000; Simon-Rojo *et al.* 2015). The key benefits include contributing to employment and the development of small-scale rural entrepreneurs; improved education and health; and to social inclusion, through integrating those at risk of social exclusion, such as migrants.

A food system encompasses the full value chain of producing food for human consumption, from agricultural activities and other means, through transportation, handling, processing, storage, distribution and consumption, to organic – including human – waste management and disposal/ reintroduction into productive use (Ericksen 2008). A food system gathers all the elements: people, environment, infrastructures, inputs, processes, institutions and activities that relate to the production, processing, distribution, preparation and consumption of food, and the outputs of these activities, including socio-economic and environmental outcomes (Figure 1).

Table 1 | Definitions of urban agriculture, urban farming and urban gardening

Definition	Source
Urban and peri-urban agriculture (UPA) can be defined as the growing of plants and the raising of animals within and around cities. Urban and peri-urban agriculture provides food products from different types of crops (grains, root crops, vegetables, mushrooms, fruits), animals (poultry, rabbits, goats, sheep, cattle, pigs, guinea pigs, fish, etc.) as well as non-food products (e.g. aromatic and medicinal herbs, ornamental plants, tree products). UPA includes trees managed for producing fruit and fuelwood, as well as tree systems integrated and managed with crops (agroforestry) and small-scale aquaculture.	FAO (2019)
Urban and peri-urban agriculture can be defined as the growing, processing and distribution of food and other products through plant cultivation and (seldom) raising livestock in and around cities for feeding local populations	Game & Primus (2015)
Urban agriculture spans all actors, communities, activities, places and economies that focus on biological production in a spatial context which – according to local standards – is categorised as 'urban'. Urban agriculture takes place in intra- and peri-urban areas and one of its key characteristics is that it is more deeply integrated into the urban system compared to other agriculture. Urban agriculture is structurally embedded in the urban fabric; it is integrated into the social and cultural life, economics and the metabolism of the city.	Vejre *et al.* (2015)
Urban agriculture is growing, processing and distribution of food or livestock within and around urban centres with the goal of generating income.	McEldowney (2017); Roggema (2016)
Urban agriculture is an industry that produces, processes and markets food and fuel, largely in response to the daily demand of consumers within a town, city or metropolis, on land and water dispersed throughout the urban and peri-urban area, applying intensive production methods, using and reusing natural resources and urban wastes, to yield a diversity of crops and livestock.	Smit (1996)
Urban agriculture is an industry located within (intra-urban) or on the fringe (peri-urban) of a town, a city or a metropolis, which grows and raises, processes and distributes a diversity of food and non-food products, (re-)using largely human and material resources, products and services found in and around that urban area, and in turn supplying human and material resources, products and services largely to that urban area.	Mougeot (2001)
Urban food gardening encompasses agricultural activities with generally low economic dependence on the material outputs while using food production for achieving other, mostly social, goals.	Simon-Rojo *et al.* (2015)
Urban farming refers to intentional business models taking advantage of proximity to the city by offering local or regional agricultural products or services. The importance of the production in proportion to the other societal benefits can vary strongly (…), both, the production-oriented side or the co-benefit-oriented side may prevail depending on the individual practices of an urban farming operation.	Pölling *et al.* (2015)

Continuous productive urban landscapes (CPUL) and growing space typologies

Agriculture is space and time-bound: plants and animals grow in a certain place and at a certain pace. Thus finding space in cities and city-regions is a prerequisite for urban farming. While some of these spaces are well known, such as traditional farming land, allotment gardens or family gardens, other spaces might offer the potential for urban farming in a permanent or a temporar way (Table 2).

The *Continuous Productive Urban Landscapes (CPUL City)* concept (Viljoen *et al.* 2005) describes an urban future based on the planned and designed introduction interconnected urban landscapes defined by urban agriculture into existing and emerging cities. It follows a systemic approach using quantifiable and qualitative arguments to propose that urban agriculture contributes to more sustainable and resilient food systems while also adding beneficially to the spatial and social quality of the urban realm (Bohn 2016). A CPUL aims to interconnect urban food-producing landscapes within a city and to the citizens on the one hand and to connect these landscapes to the rural hinterland on the other and thereby facilitates activities across all parts of the urban food system.

This was illustrated also by the outcomes of the studies that evaluated the stakeholder perceptions of rooftop agriculture in Berlin and Barcelona (Sanyé-Mengual *et al.* 2015a; Specht *et al.* 2015a, 2015b) and concluded that even if there are a number of potential risks assossiated with the urban

doi: 10.2166/bgs.2019.931

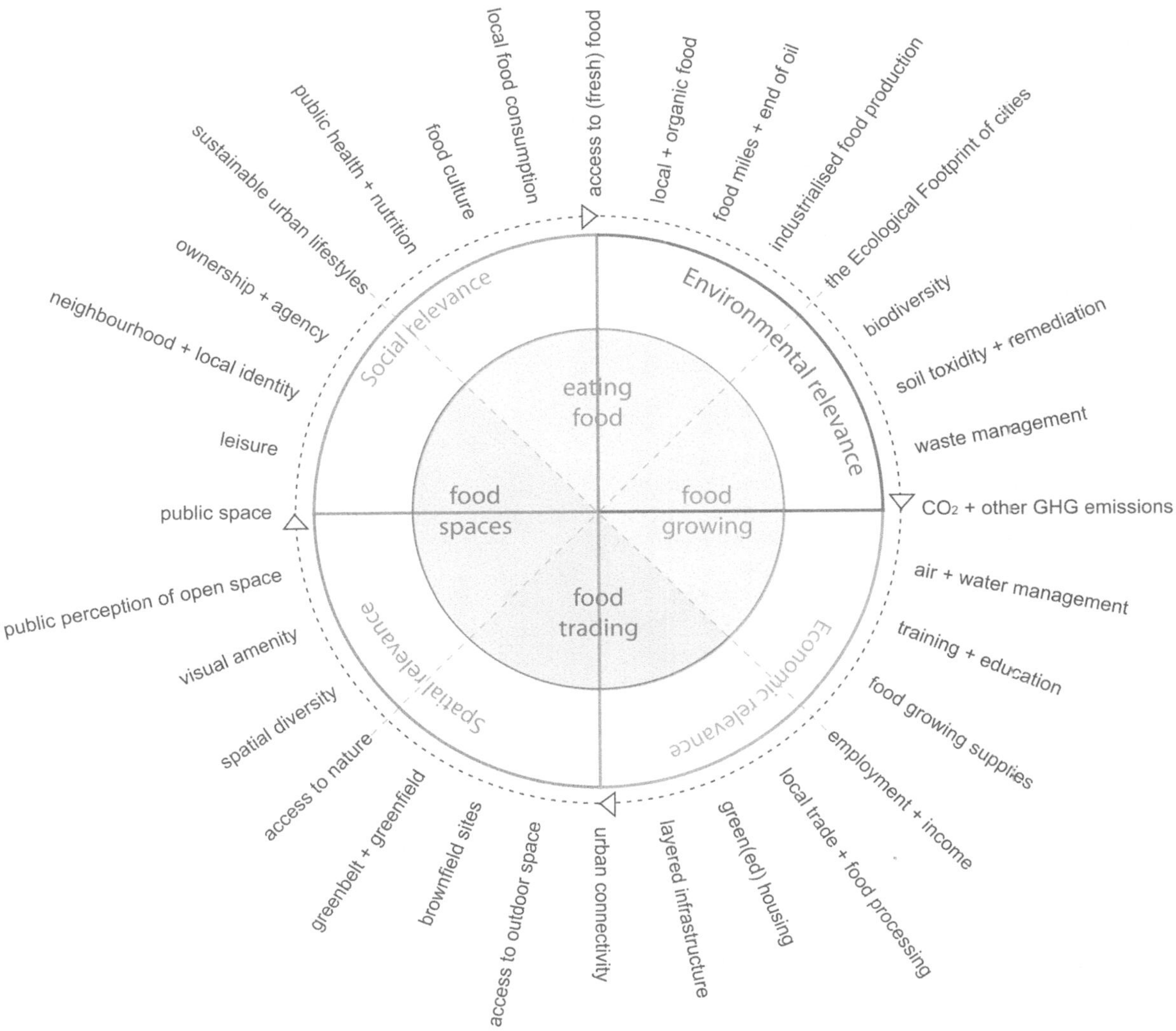

Figure 1 | This diagram shows the many links food, and the food system has with most aspects of urban life. It gives weight to spatial aspects related to urban food, including those related to urban agriculture.

farming system, stakeholders establish new market structures (e.g., short supply-chains) to overcome barriers and to ensure a socially accepted development of this new form of urban agriculture.

Relationship between urban agriculture and green infrastructure

The significance of urban agriculture has been highlighted by a set of UN-Habitat reports on how cities can work with nature. In these reports, it is argued that to achieve environmental and economic resilience; biodiversity needs to be reinstated in urbanised areas (UN Habitat 2012). One of the major co-benefits of urban agriculture lies in its contribution to the urban environment, green infrastructure and the related ecosystem services (Viljoen *et al.* 2005; Santo *et al.* 2016; McEldowney 2017; Samson *et al.* 2017; Golden & Hoghooghi 2018; Piorr *et al.* 2018).

Green infrastructure is a significant element in European planning policies on all scale levels. It is one of the primary tools for achieving the EU Biodiversity Strategy 2020 (European Commission 2011) as well as smart, sustainable and inclusive growth defined by the Europe 2020 Strategy (European Commission 2010). Urban agriculture contributes to the ecosystem services of green infrastructure as a provisioning service for food, energy and raw materials, as well as through a range of other ecosystem services (Table 3).

Table 2 | A typology of growing spaces for urban farming (adapted after Santo *et al.* 2016; Simon-Rojo *et al.* 2015)

Typology of the city area	Urban farming spaces in cities	Examples and references
Soil-bound spaces	Arable land	
	Allotment gardens	Individual gardening (commune patch)
	Family gardens	Individual gardening (private patch)
	Squatter gardens	
	Community gardens	
	Parks and other public green spaces	Van Leeuwen *et al.* (2010)
	Urban derelict land	Brown *et al.* (2016)
	Guerrilla gardening	Manchester, UK*
Mobile and soil-independent systems	Growing boxes and bags	
	Mobile containers	e.g. PAFF Box (Delaide *et al.* 2017)
Building bound spaces	Rooftops:	Buehler & Junge (2016) Glasshouses
	– Open rooftops	
	– Covered rooftops	
	– Flat roof	
	– Roof with inclination	
	Facades:	Glasshouse
	– Open facades	
	– Covered facades	
	Building extensions:	
	– Balconies	
	– Window sills	
	Indoor spaces with/without artificial lighting	
Water bound spaces	Urban streams	
	Urban stagnant waters	e.g. ponds, lakes
	Amphibia systems	e.g. floating islands

*http://www.urbanallotments.eu/fileadmin/uag/media/STSM/Binder_shortreportSTSM_final.pdf.

Tóth & Timpe (2017) have shown the quantitative importance of urban agriculture in the green infrastructure networks of four differently structured European urban regions (Dublin/Ireland, Sofia/Bulgaria, Ruhr Metropolis/Germany and Geneva/Switzerland). The regional green infrastructure systems of urban regions, which form a continuous system of open spaces often described as spatially coherent figures (green corridors, green wedges or green belts), can, in most cases, only achieve this claim of continuity if they include the semi-natural areas used for urban agriculture. Urban agriculture is an indispensable component of green urban systems.

Biodiversity and ecosystem services

Due to their population density, urban areas have a very high demand for multiple ecosystem services. Moreover, cities play an essential role in climate change mitigation and is increasingly vulnerable to climate change impacts (Rosenzweig *et al.* 2010). To achieve well-functioning circular cities that ensure good quality of life for their residents, it is urgent to safeguard biodiversity and improve the supply of regulating, cultural and supporting ecosystem services (McPhearson *et al.* 2015). Environmental benefits associated with urban agriculture include increased biodiversity, mitigation of the 'urban heat-island effect' and a reduced risk of flooding (McEldowney 2017). Urban agriculture has enormous potential to provide multiple ecosystem services in addition to food production, significantly contribute to the functioning of green/blue infrastructure and mitigate climate change (Lwasa *et al.* 2014). However, to maximise multiple benefits that can arise from urban food production, urban agriculture has to adopt sustainable farming practices (for instance organic farming, use of agroecological approaches), ensure functional integration to the urban fabric, and

doi: 10.2166/bgs.2019.931

Table 3 | Contributions of urban agriculture as green infrastructure (GI) based on Timpe *et al.* (2015)

Ecosystem Services TEEB (2015)[a]: ES Benefits Group	European Commission (2013a, 2013b): GI Benefits	Possible contribution of urban agriculture
Provisioning services	Multifunctional resilient agriculture and forestry	Providing food, fibre and biomass and enhancing pollination.
	Investment and employment	Employment in agriculture, investment in agricultural enterprises and buildings, productive and maintained land as contributions to a better local image.
Cultural services	Tourism and recreation	A broad range of recreational activities proposed on farms and in gardening associations, farms and gardens as a destination.
	Education	Agriculture as a teaching resource and 'natural laboratory'.
	Health and wellbeing	Farm work and gardening as activities for physical and mental health, access to healthy local food.
Regulation services	Enhanced efficiency of natural resources	Maintenance of agricultural soil fertility, pollination through urban beekeeping.
	Climate change mitigation and adaptation	The cooling effect of agricultural areas, carbon storage in soils.
	Water management	Groundwater recharge and purification under agricultural soils, stormwater retention.
	Land and soil management	Reduction of soil erosion, maintaining/enhancing soil's organic matter, increasing soil fertility and productivity, mitigating land consumption, fragmentation and soil sealing.
	Disaster prevention	Flood hazard reduction through stormwater retention and agricultural polders, erosion control.
Habitat	Conservation benefits	Maintenance of agrobiodiversity, maintenance of agricultural habitats.
	Low-carbon	Short-chain food provision, local bioenergy from agriculture.

[a]The Economics of Ecosystems and Biodiversity – TEEB.

safeguard biodiversity key areas (IPBES 2019). This applies to both small-scale food production for personal or community use (urban food gardening) as well as commercial farming in urban areas (urban farming). Table 3 shows the contributions of urban agriculture as this is what the green infrastructure is based on.

Air pollution risk and agronomic considerations on urban farming

Pollution risks of urban food farming, understanding food production outsides in open city spaces, can be combined with risk for food safety and content of pollutants in the food products. There are a few articles on this topic but more research is needed to fully understand the risks. There are three primary risks of gardening in cities and urban environment: soil, water and air pollution. For air pollution, there are three categories: (1) not accumulated in plants, (2) transport-vectors of pollutants and (3) pollutants that are taken up in plants. Within a city, many sources of pollution are present: e.g. traffic, industries, heating (Ortolo 2017).

Fruit trees are most affected by air pollution. The air pollution in China has caused damage to fruit trees by delaying sprouting, shortening the flowering period, accelerating senescence and reducing CO_2 assimilation. This resulted in a reduction of fruit numbers and premature dropping of fruit (Zheng & Chen 1991). High ozone concentrations cause chlorotic spotting, necrotic lesions and premature senescence in trees, vegetable crops and cereals (Rich 1964; Krupa *et al.* 2001). The ambient air pollutants (SO_2, NO_x, SPM and RSPM) caused a significant reduction in total chlorophyll, carotenoid, ascorbic acid, plant height, shoot fresh weight, root fresh weight and yield of wheat and mustard crops grown at polluted sites (Chauhan & Joshi 2010). The elevation of CO_2 concentration has been shown to increase the yield of crops under laboratory conditions, but in reality the degree of growth stimulation is dampered in the environment due to high temperatures and increasing tropospheric O_3 (Ainsworth 2008). A case study in Varanasi, India concluded that gaseous pollutants such as SO_2, NO_2 and O_3 have damaging effects on the yield of wheat, mustard, mung and palak plants (Agrawal *et al.* 2003).

Apart from ozone, it can be difficult to identify the causal link between a specific gas and damage to crops, which is why there are few studies made on this topic (Hamel & Danso 2011). A study in Greece used a combined air quality and GIS modelling approach to estimate crop damages from photochemical air pollution (O_3) and depict the corresponding economic damages. Total economic damage to crops turned out to be significant and estimated to be approximately 43 M€ for the reference year (Vlachokostas *et al.* 2010).

Suspended particulate matter has the most significant effect on crop yields and the quality of the crops. Atmospheric pollutant deposition has been noted as the most common pathway for lead contamination of leafy greens in Uganda (Nabulo *et al.* 2006). A study in Nigeria of the correlation between traffic emissions of Cd, Cu, Cr, Ni, Pb and Zn near a highway and the concentration of heavy metals in the vegetation and soil samples near the highway has been conducted. The study showed that roads have a significant effect on heavy metal accumulation in vegetation (Ndiokwere 1984).

MATERIAL AND RESOURCE MANAGEMENT

Resource management

Resources such as water, food and energy are what cities need to function (Figure 2). The entirety of resource streams and their interactions with each other, citizens and urban space are incredibly complex. Despite some of these resources being nature-based, most resources flows are linear: they are used and then disposed of. This linear urban metabolism (Rogers 1997; Daigger 2009) dominates in contemporary towns and cities all over the world, and it is causing a plethora of problems. The

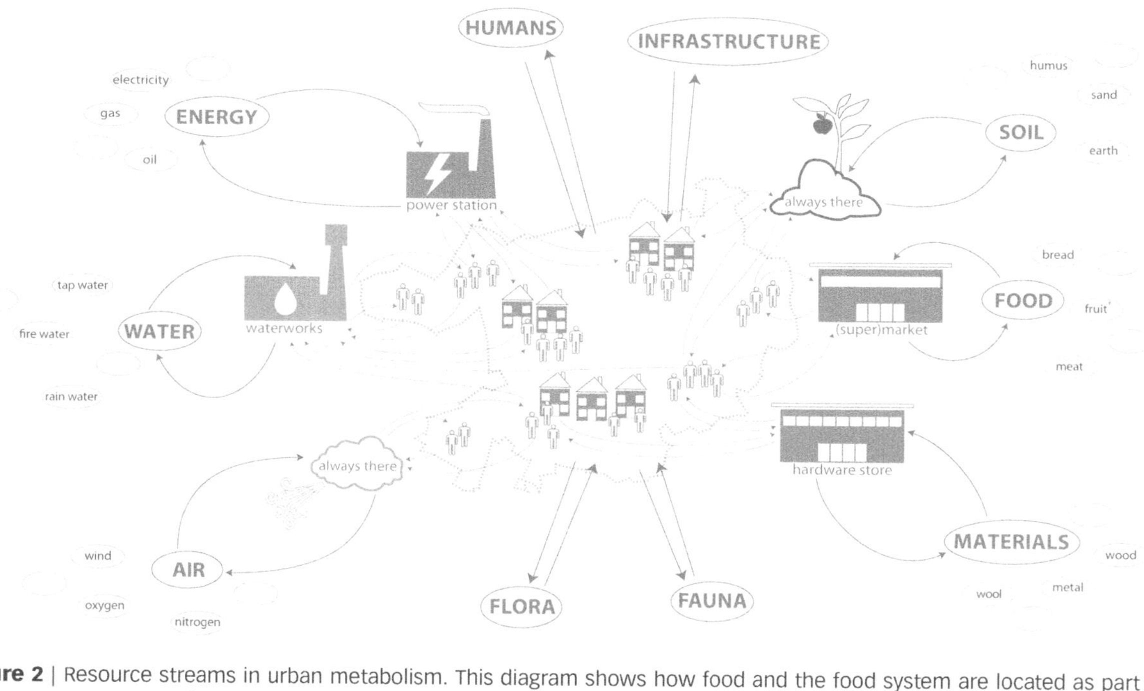

Figure 2 | Resource streams in urban metabolism. This diagram shows how food and the food system are located as part of other resource streams into and out of the city (*Source*: Bohn 2016).

doi: 10.2166/bgs.2019.931

circular approach explores how to manage resources by reducing, reusing and recycling. As a consequence, this leads to redesign of the urban, peri-urban and rural space, and to a new conceptualisation of their interlinkages. One of the critical elements in the development of closed-loop designs is a localised food systems approach, which also links to a broader understanding of a (peri-)urban water–energy–food nexus (Figure 2).

Growing media

Soil is the 'default' growing substrate for plants, however urban soils are often degraded (De Kimpe & Morel 2000) and do not enable healthy plant growth. Therefore, replacing the soil with other substrates in urban environments, can contribute towards healthy produce. Based on chemical properties, growing media can be split into organic and inorganic (Table 4). The most important physical property is particle size, which affects the physical characteristics (e.g. porosity, water holding capacity, air space). Well-balanced physical and chemical properties of the growing media induce the plant growth and also promote biological activity (Maucieri *et al.* 2019). Also, it should be free of weeds, pathogens and toxins.

In circular cities of the future, it is important to focus on renewable materials from agricultural, industrial and municipal waste streams to identify beneficial and environmentally sustainable materials as growing media (European Commission 2019). It is environmentally beneficial to reuse and recycle renewable materials, and contributing to the circular economy (CE) which is, together with bioeconomy, supporting concepts in order to facilitate the transition to a sustainable society.

Deposition of pollutants in soil

Particulate air pollutants are usually settleable by gravity and are deposited on the ground through wet and dry deposition. They cause acidification, salinisation and high heavy metals concentrations. To assess the influence of air pollution on soil composition, a study of heavy metals concentrations (Cd, Pb, Ni, Sb and Bi) in the settleable particulate matter in two locations in Spain has been performed. The study showed significant seasonal variability for heavy metal content and a strong dependence on rainfall in the area. The maximum values of heavy metals were measured in spring or autumn when there was the highest rainfall (Soriano *et al.* 2012).

Werkenthin *et al.* (2014) reviewed studies of metals in European roadside soils and concluded that the highest levels of Cr, Cu, Ni, Pb and Zn, were determined in the topsoil layer, and located in the first 5 m beside the road. Generally, the influence of traffic on soil contamination decreased with increasing soil depth and distance to the road.

Based on these findings, there are some concerns about the quality of food produced in the urban environment. The suitability of food produced in close proximity of urban traffic or other sources of pollution, should be closely examined.

Available water resources

Treated domestic or municipal wastewater, also designated as reclaimed water, is used worldwide as an alternative water source. In some countries is even used as a water source for drinking water, such as in Singapore and Texas-USA (Yi *et al.* 2011). In California-USA, in full-scale large dimension projects, reclaimed water is being used for irrigation. In the near future, when the health and social objections existing presently in Europe, are overcome by the overwhelming current efficiency of advanced wastewater and water treatment technology, reclaimed water will become one of the most important sources of urban water (Norton-Brandão *et al.* 2013).

Table 4 | Overview of substrates used in urban agriculture

Substrate	Description	References
Soil		
Urban soil	Urban soil is often exposed to many strong influences which result in contamination and structural deterioration. Among others, urban soils can be contaminated with hydrocarbons and/or heavy metals, which can accumulate in produce and compromise human health. Before growing food in urban soils, soils have to be tested for exceeding the contaminant limits. Contaminated soil can be remediated with physical (soil excavation, washing and vapour) and biological (microbial, fungal remediation, phytoremediation) techniques.	Clarke *et al.* (2015); Jean-Soro *et al.* (2015); Schwarz *et al.* (2012)
Substrates of organic origin		
Peats	Result from anaerobic decomposition of peat mosses under waterlogged conditions. Depending on conditions under which they were generated, peats possess superb physical, chemical and biological properties suitable for plant growth. However, peat bogs are an endangered ecosystem and are mostly under protection. Therefore, peat substitutes should be used preferentially.	Krucker *et al.* (2010); Michel (2010); Maucieri *et al.* (2019)
Peat alternatives	Motivated by environmental, but also technical issues, many peat alternatives have been investigated, for example, ground fresh rice hulls (GRH), anaerobic digestion residues (ADR), coir dust or cocopeat, wood fibre substrates. Some have excellent properties to be used in soilless culture for the production of seedlings and transplants.	Schmilewski (2008); Zanin *et al.* (2012); Gruda & Schnitzler (2006)
Compost	The result from the aerobic decomposition of plant material. When mature, composts ensure minimal medium shrinkage, oxygen consumption, nitrogen immobilisation and phytotoxicity. A wide array of organic waste can provide feedstocks for composting, this being the reason why it can be widely implemented also on a household level and in the cities.	Maher *et al.* (2008); NiChualain *et al.* (2011); Raviv (2013); Barrett *et al.* (2016); Perez-Murcia *et al.* (2006)
Vermicompost	Results from the composting process using various species of earthworms and organic materials, such as plant and food waste.	Bachman & Metzger (2008)
Pyrolysis biochar	Biochar resulting from the pyrolysis of organic matter. It can be used as a soil amendment or as a part of substrate mixture. Pyrolysis process requires predominantly dry substrates (e.g. straw, faeces, wood chips).	Bruun *et al.* (2012); Gold *et al.* (2018)
HTC (hydrothermal carbonisation) biochar	Biochar resulting from hydrothermal carbonisation of organic matter. HTC also functions with predominantly wet substrates like sewage sludge or whey. It can be used as a soil amendment.	Escala *et al.* (2012)
Inorganic media		
Sand and gravel	The coarse fractions of the soil minerals have particle size 0.02–2.0 mm (sand), and 5–20 mm (gravel). Coarse sand is preferred as a substrate for plant growth and rooting cuttings while fine sand is preferred for seedling production. Quartz (SiO_2) is the most common component of the sand fractions. Due to environmental constraints, natural sand dune extractions limit the use of sand as a growing media. Gravel is less used due to the low water holding capacity and heavyweight.	Lennard & Leonard (2006)

(*continued*)

Table 4 | continued

Substrate	Description	References
Perlite	A volcanic based inert, lightweight mineral with high porosity. It is produced at temperatures above 1,000 °C. It has a pH of 7.0–7.5 and contains no minerals available for plant needs. It is produced in various particle sizes. It is used in mixtures with other media. Similar to perlite, pumice is another volcanic based material.	Verdonck *et al.* (1981); Maucieri *et al.* (2019)
Zeolites	Are usually formed by the metamorphosis of volcanic rocks but also from non-volcanic materials in marine deposits or aqueous environments. They have high ion exchange, adsorption, hydration-dehydration and catalysis properties, therefore also high pollutant removal capacity. Zeolite is used as a growing media component.	Ming & Mumpton (1989); Harland *et al.* (1999)
Vermiculite	A natural clay mineral with water molecules within its structure layers. It is produced similarly to and has similar physicochemical properties as perlite. It is produced in several particle sizes, which affects the physical characteristic of the material (e.g. porosity, water holding capacity, air space). Vermiculite has a pH of 7.0–7.5, low electrical conductivity (EC), and contains potassium (K) and magnesium (Mg).	Verdonck *et al.* (1981); Maucieri *et al.* (2019)
Mineral wool/ Rockwool	It is industrially produced by melting minerals at up to 1,600 °C and spinning the molten mixture at high speed into thin fibres (~5 μm diameter). Rockwool is often used in soilless cultures, providing advantages (sterile, inert and consistent in performance) but also limitations (lacks nutrient buffering capacity) in its use. However, it is a non-renewable resource, and the possibilities for recycling are currently limited.	Verdonck *et al.* (1981); Maucieri *et al.* (2019)
Nano-fertilisers	Nanotechnology can be used in the production of fertilisers due to the high efficiency and the homogenous distribution of nano-form of the nutrients. Nanoparticles (1–100 nm) confer improved efficacy in their physicochemical properties. However, the plants response to the nano-fertilisers is significantly different and dependent on plant type.	Torabian *et al.* (2017); Nair *et al.* (2010)
Light expanded clay aggregate (LECA)	LECA (ISO 10–20) is a building material made of clay, burnt and converted into small, porous, hard-surface spheres. The balls are used in growing beds for plants, as insulation, and as raw materials for blocks, pipes and other elements.	Maucieri *et al.* (2019)

Linking urban water usage to urban agriculture has the potential to be mutually beneficial. Availability of safe alternative water sources may: (i) facilitate higher uptake of urban agriculture, (ii) proper use or reuse of municipal water which may improve stormwater and wastewater management, reduce sewer overload and nutrient loads to urban rivers and allow sewer mining for resource recovery (e.g. nutrients) (Tahir *et al.* 2018; Voulvoulis 2018). However, any long-term studies on the combined benefits, health risks and robustness of reuse systems in farming projects appear to be missing and are the research focus of only very few current projects, such as the case of Roof Water Gardens project (Million *et al.* 2016) and the HOUSEFUL project on innovative circular solutions and services for the housing sector (www.houseful.eu).

Urban agriculture in a circular city should meet its water requirements by water resources which originate from within the urban watershed (Fletcher *et al.* 2013; Tahir *et al.* 2018; Voulvoulis 2018; Pratt *et al.* 2019); and, in this framework, tap water should not be the first choice. More appropriate resources may comprise natural rainfall for rainfed farming, the usage of rainwater temporarily stored in cisterns – also called rainwater harvesting – or the usage of urban wastewater. Untreated urban wastewater is usually not considered in farming projects in developed countries due to significant public health concerns for farmers and consumers (Khalil & Kakar 2011; Drechsel *et al.* 2015; Khan *et al.* 2015; Okorogbona *et al.* 2018). However, the usage of treated or untreated greywater – wastewater generated in households or office buildings from streams without fecal contamination – (i.e. all streams except for the wastewater from toilets) recently received more attention as it also reclaims fertiliser resources such as nitrogen, potassium, calcium, magnesium, sodium and phosphorus (non-renewable resource) (Qadir *et al.* 2007; Chen *et al.* 2013; Oteng-Peprah *et al.* 2018). Some studies have shown that nutrient-rich wastewater can be productively reused in urban and peri-urban agricultural systems, contributing to crop yield and improving soil fertility, thus enhancing the resilience of urban areas (Murray & Buckley 2010; Drechsel *et al.* 2015).

Irrigation water requirements

Some projects – including Urban GreenUP (www.urbangreeup.eu), CITYFOOD (www.cityfood.igb-berlin.de) and TUNESinURB (www.tunsinurb.org) – have studied the sustainability of gardens, edible gardens and urban farming (Paço *et al.* 2019). The knowledge on irrigation water requirements considers a scenario where rainwater is harvested and stored for the irrigation season as a possible alternative or supplement to the current irrigation sources in Mediterranean cities, mainly where rainfall is less than 500 mm/year, concentrated in some months in wintertime, and the current water source is tap water. However, technical solutions to accommodate the water volumes involved, in what regards building structure and architecture, are needed as well as water pumping and irrigation systems costs analysis. Urban landscapes for environments with hot/dry summers can benefit from the use of low water requirements plants, namely native species (Paço *et al.* 2019). It is essential to quantify the water requirements of such species as little information exists.

CULTIVATION TECHNIQUES AND PRODUCTION SYSTEMS

The lack of soil fertility, available rural and urban land space, agricultural sectors' long distances from the urban centres provide the challenges and opportunities to develop urban agriculture. Therefore, in addition to conventional growing techniques in soil, different soilless techniques are implemented.

Soilless cultures and hydroponics

Soilless culture is a technique to grow plants without soil, using inert media (e.g. rockwool, clay pebbles, coconut fibres) or no media, and supplied with a nutrient solution (i.e. water and soluble nutrients). Currently, the terms soilless culture and hydroponics are used as synonyms; however, hydroponics originally meant cultivation in a nutrient solution without supporting growing substrate as soil (Zanin *et al.* 2009).

In the horticulture, soilless cultures are the most important cultivation methods for effective production in greenhouses (Jensen 2010). The criteria for classification of soilless cultures are: presence and properties of substrates and containers, vertical or horizontal system, location (greenhouse, garden, integrated into the building), how the nutrient solution is administered to the plant (dripping watering, immersing in stagnant solution or through mist spray), and type of water circulation (open or closed systems) (see also Maucieri *et al.* 2018a, 2018b; Maucieri *et al.* 2019). Among the newest hydroponic technologies are 'aeroponic systems', drip irrigation and nutrient film technique (NFT). The most frequently cultivated species in this type of culture are vegetables, herbs and

doi: 10.2166/bgs.2019.931

medicinal plants. Under suitable conditions, decorative plants (e.g. roses, gerberas, carnations) can be grown as well (Savvas & Passam 2002).

Aquaponics

To succeed with integrated production units, producing more than one type of product for sale, are highly sought. Aquaponics is a production technology which combines aquaculture production in recirculating aquaculture systems (RAS), with the soilless cultivation of plants (Graber & Junge 2009; Rakocy 2012; Monsees *et al.* 2017). The effluent from the fish (or other aquatic organisms) production unit supplies the horticultural unit with water and nutrients for plant growth. Since the nutrient profile can be individually adjusted by measuring the nutrient profile and adding missing nutrients, multiple plant species can be grown as monocultures or in polycultures (e.g. intercropping, companion planting (Maucieri *et al.* 2017)). A wide array of vegetables (Graber & Junge 2009; Monsees *et al.* 2019), flowers (Agha Rokh 2008), fruits (Schmautz *et al.* 2016), herbs (Nozzi *et al.* 2018) and berries (Villarroel *et al.* 2011) can be produced and serve the local market. Pest and disease management focuses on prevention and is based on principles of integrated pest management and organic agriculture (Némethy *et al.* 2016). Very different system set-ups can be customised to diverse requirements; high-/low-technological, commercial sizes, backyards systems, education and hobbies set-ups (Maucieri *et al.* 2018a, 2018b). Most common are freshwater systems on-land (Skar *et al.* 2015).

Vertical farming

Vertical farming is a system of farming whereby living organisms (animals, plants, fungi and other forms of life) that are cultivated for food, fuel, fibre and other products or services are artificially stacked above each other, vertically. The concept of vertical farming is integrated into the urban production of fresh produce. These systems are very efficient in terms of land use due to reduced dependency on land resources (Pérez-Urrestarazu *et al.* 2015). Moreover, vertical farming can contribute to the effectiveness of the arable area for crops by constructing a high-rise building with many levels on the same footprint of land (Despommier 2010). Soilless culture and hydroponics can add inputs to that direction, with considerable savings on water, minerals and phytochemicals through the sustainable cultivation cropping systems.

POLICIES AND REGULATIONS

Governance

The governance of UA may primarily include such issues as land, land use, access, food and ecosystem health, education and the environment as well as heritage and cultural practices (Corcoran & Calvin 2015). Prové *et al.* (2015) established a conceptual framework for urban agriculture governance processes and identified characteristics which influence the processes of management of urban agriculture initiatives. The three levels of this framework, which include the main features of the governance of urban agriculture are: (i) **Urban context** (including the local geographic situation, economic and political situation, the agricultural context and the status of urban-rural relations); (ii) **External governance characteristics** (including public policies, partnerships, legitimation processes); and (iii) **Internal governance characteristics**, which include the project objectives, spatial scale, temporality, actors and resources (land, finance and knowledge mobilised in the project). All these are embedded into the local situation, characterised by geography, climate, economic and political situation, cultural values and urban–rural relationships.

Of course, public policies that influence these three categories are very different (Figure 3). For example, urban gardening is not affected by agricultural policy. Thus, the analysis of governance will focus on urban agriculture initiatives with the active involvement of professional farmers and public policies that influence these initiatives, mainly agricultural policy and planning. The focus will be on the integration of agriculture and agrarian actors in the development of cities and especially in regional planning (UNUIAS 2010).

There are huge variety of UA activities and interventions, which can involve domestic, public and commercial projects, involving actors of different resources, skills, orientations and motivations. In general, because of intrinsic hybridity of UA governance, the policy linkages are often unexplicit and not considered strategical either by national governments or municipalities (Corcoran & Calvin 2015).

BUSINESS MODELS IN URBAN AGRICULTURE

Economically viable urban food production

City environments influence agriculture (Heimlich & Barnard 1992). Farmers located in such areas will have to adjust their farming to exploit all opportunities and to counter most of the restraints (van Huylenbroeck *et al.* 2005). The main hurdles for urban agriculture can be summarised into land-related constraints, conflicts caused by improper behaviours of urban dwellers, and economic incentives within cities outside farming. However, urban areas hold chances for economically viable food production strategies. Little attention was given to these positive effects of cities and agglomerations on farming in the past (Beauchesne & Bryant 1999). Cultivation, processing and marketing of urban farming's food and non-food products take place in an environment of the highest demands (McClintock 2010). The potential of nearby and easily accessible large consumer groups, the concentration of particular societal demands and trends, and the innovative milieu in cities offer favourable framework conditions for local and short marketing channels both for agricultural products and for the provision of services associated with farming.

Urban influence on business performance and success

Cities and agglomerations increasingly incentivise farms to adjust to the urban conditions aiming to achieve profitability and business success. By doing so, farms increase chances to maintain economically viable or enhance their business performance (van Veenhuizen & Danso 2007). When farms do not adjust adequately to the multifaceted and dynamic urban influences, they increasingly tend to give up

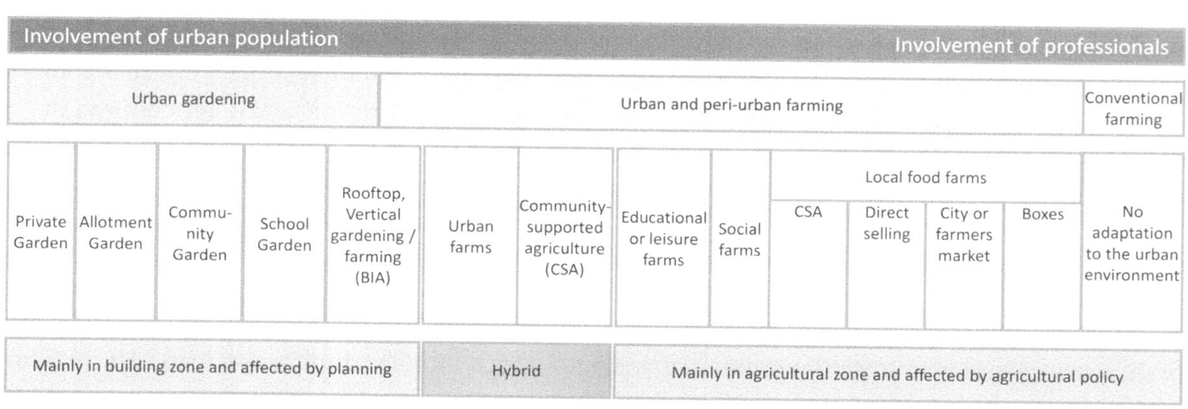

Figure 3 | Typologies and social aspects of urban agriculture initiatives (adapted after Mumenthaler 2015).

doi: 10.2166/bgs.2019.931

or turn into part-time or hobby farming with main revenues originating outside of agriculture (Zasada 2011). Gardner (1994) says that commercial farming in urban areas is surviving and even prospering when adjusting adequately to the cities. Thus, the diversity and complexity of urban influences result in a variety of city-adjusted farm strategies. Urban agriculture has been identified as being more diversified, polarised and multifaceted than elsewhere (Zasada 2011). Common strategies of urban agriculture focus on high-value production, product niches, short supply chains, Alternative Food Networks (AFNs) and the provision of services connected with agriculture (e.g. Heimlich & Barnard 1992; Gardner 1994; Bailey *et al.* 2000; Mougeot 2000; Houston 2005; Zasada 2011; Aubry *et al.* 2012; Aubry & Kebir 2013). Specialisation, niche production, multifunctionality, food chain management, quality of food, and embeddedness of food are listed by Wästfelt & Zhang (2016) as appropriate for urban agriculture activities. By focusing on the consumer side, Barbieri & Mahoney (2009) and Inwood & Sharp (2012) highlight that better chances of farm business survival and development exist for those city-adjustments which apply immediate consumer orientations and relationships. Agricultural innovations often take place on farms within metropolitan areas and subsequently diffuse into rural areas (Prain & de Zeeuw 2007; Elgåker & Wilton 2008; Zasada 2011).

Business model classifications

The heterogeneity of urban farming's city adjustment strategies, as well as the lack of business model, approaches are highlighted by Boons & Lüdeke-Freund (2013). Both have been providing the basis for the recent emergence of business model classifications in urban agriculture since a few years. While economies of scale is still an essential rural farming business model to stay competitive under intense cost pressures in the food sector, urban agriculture business models have to distinguish by adjusting to cities and move away from mainstream commodity market and global prices mechanisms. New business concepts have emerged on established (peri-)urban farms and also by newcomers and start-ups in urban agriculture. The specifically challenging, but also enabling urban conditions encourage innovations in farming, and result in the appearance of business models, which in many respects are different from rural farms. Product differentiation and enterprise diversification are the prevailing business models, but new forms of and new actors in urban agriculture raise experimental, shared economy and experience to emerging business models (Pölling *et al.* 2015; van der Schans *et al.* 2015).

The business model differentiation is frequently applied in urban areas to create distinctions in production, processing and marketing from the bulk market. Short food chains, especially direct sale, along with premium prices for specific product features (for instance super-fresh, ethnic, tasteful) are based on personal, transparent and honest producer/consumer relationships. Cost reduction represents the business model closest to rural farming. However, also farms located in agglomerations' peri-urban fringes use this low-cost approach for profitability, and in the urban context, specific expressions have emerged. Commons are specialisation in high-value crops (horticulture) and methods to reduce costs, like using available and cheap urban surplus resources (heat, sewage water, biomass) (Pölling *et al.* 2015).

The business model: sharing economy model

Lately initiatives based on 'sharing economy' (or 'the commons') increasingly gain importance as an expression of the new economy. Resources required to run urban agriculture in the form of a shared economy model, e.g. Community Supported Agriculture, are jointly mobilised and managed: land, labour, credit, tools, machinery, network contacts and knowledge. The experience focuses on providing authentic and catchy memories by selling a story (experience) in addition to a product. Place-making and training or leisure activities are essential elements that in this model are combined with food production.

ANALYSIS OF CURRENT EUROPEAN PROJECTS

Methods

The publications and projects were analysed regarding the potential contribution of the elements of urban farming systems to circular economy approach for a resourceful, resilient and sustainable city.

This study was based on the literature available from various bibliographic reference databases (namely Google Scholar, Research Gate, Web of Science, Scopus) and the work of CA17133 members. The systematic literature review process was applied to select the latest and the most relevant studies on this particular topic. Specific terms were used, like 'nature-based solutions in cities and vegetable production' and synonyms for 'urban agriculture', 'urban farming', 'closed circular systems', 'business models'. The article's relevance was discussed and grouped into three to six persons (authors) per groups, dealing with each topic related to urban agriculture as a keystone contribution towards securing sustainable and healthy development for cities in the future.

Defining all the terms, from urban agriculture to the contribution of elements of urban farming systems – specially developed for dense urban areas (e.g. underground, vertical and rooftop farming) – in a circular economy approach, helping to understand the status of a circular city based on project-database of COST Action CA17133. Two surveys were developed to the COST members concerning the: (1) projects participating in the COST Action; and (2) water sources, treatment, storage and irrigation systems. In the first survey, we combined expertise in food systems governance (Moragues-Faus *et al.* 2017), innovative production systems and business models, food safety, food waste, water consumption, irrigation systems, knowledge transfer, education, participation, alternative protein sources for feed and urban–rural nexus analysis. It was screened through the project-database and added peer-reviewed publications which we identified as key-articles for our field of research. It was also identified research gaps and extracted research questions which need to be answered to proceed into the design of a circular city. No quantitative information on current water sources, treatment, storage and operating irrigation system are available for current urban farming projects. To fill up this gap, the second survey was carried out within the 250 participants of the COST Action CA17133 Circular City, in February 2019 on water resources of urban farming research projects. (To find the nine questions, the list of surveyed research projects and their geographical distribution, please see the supplemental material.) The case of *green roofs in Lisbon* with native species study of urban landscape area, allows enhancing water use and sustainability in Mediterranean conditions.

CIRCULAR CITIES: A SURVEY OF PROJECTS PARTICIPATING IN THE COST ACTION CA17133

The relevant projects based on a keyword search through the COST Project collection are shown in Table 5. Fourteen keywords were selected based on the definitions of Food Systems (Table 1) definition and are listed on the table (see columns D to R). About our 14 words searched, several gaps where identified with significant absence: Food transportation and distribution, Food processing and transformation, Food storage, Consumption and Compost. Few references were found related to governance, consumers, citizens, and authorities. We analysed 13 projects (Table 5). However, some of them need to be seen as extremes. Project 9 (Pyrolysis of faecal wastes) is listed due to its potential to close the loop of the system.

The research areas addressed were: food policies, food systems, stakeholders engagement and awareness, urban and peri-urban agriculture for sustainable local development.

The survey includes also answers from COST members on the following questions:

1. What are the critical questions from a members point of view?
2. Critical points (gaps) enabling the contribution of urban farming to circular cities?

doi: 10.2166/bgs.2019.931

Table 5 | Relevant projects identified by searching project-database of the COST Action on Circular Cities and publications

No.	A Project	B Website if applicable	C Location
1	Integrating Edible City Solutions for social resilient and sustainably productive cities – EdiCitNet	https://cordis.europa.eu/project/rcn/216082/factsheet/en	Tygron and other world cities
2	Urban and Peri-Urban Agriculture for sustainable local development: The Multi-stakeholder Policy Action Planning as a tool for reconciling sectorial policies	http://bit.ly/UPAforSustainableDevelopmentMuiti-StakeholderPolicyAction	Portugal
3	CITYFOOD – Smart integrated multi-trophic city food production systems – water and energy-saving approach for global urbanisation (2018–2021)	https://www.igb-berlin.de/en/project/cityfood	Norway, Sweden, Germany, the Netherlands, Brazil and the USA
4	Continuous Productive Urban Landscape: Designing urban agriculture for sustainable cities	http://www.foodurbanism.org/cpuls-continuous-productive-urban-landscapes/	Worldwide (sustainable urban design concept)
5	GEOFOOD (2018–2021)	https://geofoodproject.eu/	Iceland, Slovenia and the Netherlands
6	Living from the Earth – OTKA 116219 and OTKA 100682 (2016–2019) dealing with rural-urban divide and to make rural life attractive in Hungary, the changing role of the local small-scale agri-food production	https://www.aur.edu	Hungary
7	NACHWUCHS – Nachhaltiges Agri-Urbanes zusammenWachsen (Sustainable AgriUrban Growth)	https://urbact.eu/agri-urban	Germany
8	Plattform 'Produktive Stadt' [Platform 'Productive City']	http://blogs.brighton.ac.uk/pulr/2019/06/10/platform-productive-city-holds-its-2nd-participatory-workshop-germany/	Berlin (Germany)
9	Pyrolysis of faecal wastes	https://www.zhaw.ch/de/lsfm/institute-zentren/iunr/ecological-engineering/oekotechnologie/biochar-sanitation/	Waedenswil (Switzerland)
10	Productive Green Infrastructure for post-industrial urban regeneration (proGIreg)	http://www.progireg.eu/	Aachen (Germany)
11	Urban Allotment Gardens in European Cities: Future, Challenges and Lessons Learned	https://www.cost.eu/actions/TU1201	Riga (Latvia)
12	Characterisation of nutrient recycling processes of a model aquaponic system	https://www.zhaw.ch/de/lsfm/institute-zentren/iunr/ecological-engineering/oekotechnologie/aquaponic/	Waedenswil (Switzerland)
13	The smart and sustainable city district of the future	https://www.balticurbanlab.eu/goodpractices/hiedanranta-smart-and-sustainable-city-district-future-tampere	Häme (Finland)

The search keywords were: circular cities.

3. Why is the understanding of urban farming within the food system critical to circular cities?
4. Why the understanding of urban farming within the food system is a critical shift from linear cities to circular cities?

The topics presented in the thirteen projects were (Table 6): stakeholders, food, production, transportation and distribution, storage, processing and transformation, food policy(-ies), food system, governance and authorities, consumers and citizens, farmers, urban farming, compost and the

Table 6 | The goals of relevant projects identified by searching project-database of the COST action on circular cities and publications

	Project number												
Topic	**1**	**2**	**3**	**4**	**5**	**6**	**7**	**8**	**9**	**10**	**11**	**12**	**13**
Stakeholders	x	x	x	x				x			x		
Food	x	x	x	x	x	x	x	x	x		x	x	no
Production	x		x	x		x	x	x	x	x			
Transportation/Distribution				x									
Storage				x									
Processing/Transformation				x									
Food Policy(-ies)		x	x	x		x							
Food System	x	x	x	x	x								
Governance/Authorities				x				x			x		
Consumers/Citizens	x	x		x				x					
Farmers			x	x									
Urban Farming						x	x					x	
Compost				x									
Circular city				x									x

The numbers 1–13 denote the projects listed in Table 5.

circular city. Most projects covered topics related to food (11 out of 13), production (8 out of 13) and stakeholders (6 out of 13).

Water sources, treatment, storage and irrigation systems

The survey yielded 22 research projects, which geographically covered most of Europe. Water sources used by the projects were very diverse (Figure 4).

Interestingly tap water, as the only water source was used by only 23% of the projects, while most projects (45%) used a mixture of sources. The use of wastewater, greywater or stored rainwater was rare – one project relied to 100% on natural rain.

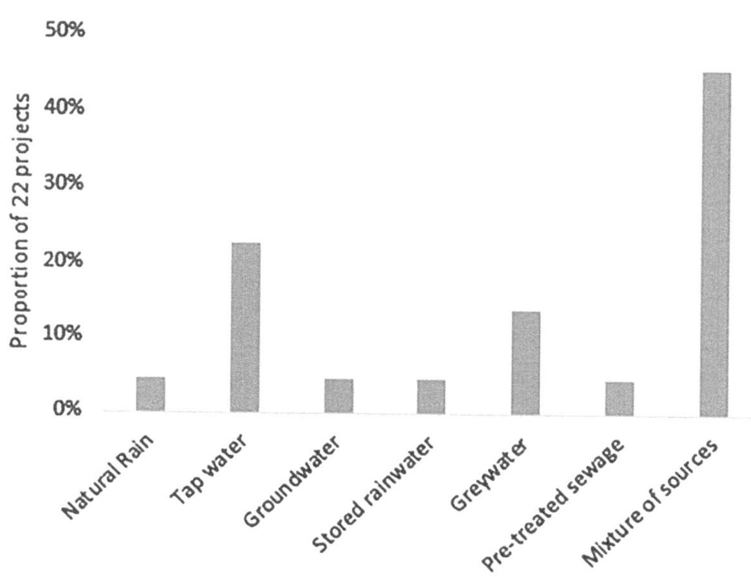

Figure 4 | Sources of water supply for 22 recent urban agriculture research projects participating in the COST Action CA17133, February 2019.

doi: 10.2166/bgs.2019.931

Regarding prior water treatment, 59% of the projects did not use any treatment of their water. The remaining used a variety of treatments including mechanical filters, plant-based, bio-reactor, sedimentation or disinfection systems. Only two of the ten hydroponic systems used greywater as a source.

The survey showed that (i) rainwater management or reuse infrastructures are often not readily in place to be used for urban agriculture projects, (ii) farming projects rely heavily on tap water and (iii) circular usage of water is not yet common except in the hydroponic projects.

CONCLUSIONS

The development of future cities is now advancing into smart and sustainable cities with a global understanding of how vital food production, supporting the circular economy, is. In 2050 about 50% of all people on earth are living in cities and will need clean water, food, energy, social space, meeting points, relax areas and knowledge pools.

Land use and planning

It is visible that municipalities around the world are beginning to consider food issues in their city planning proactively, most evident perhaps in the Food Policy Pact signed by more than 180 cities since its foundation in 2015 during the Milan World Expo. This rapid development has followed the mainly practice-based, citizen-led experiences with urban agriculture that emerged, as a conscious movement, in cities and at their edges for nearly 20 years. It is now of great importance to integrate urban food production and other food system activities into urban planning, thereby linking sustainable food provision and circular resource processes to infrastructural productive urban landscape development.

Water systems

Assessment of underlying spatial and temporal variability in water use decisions at a landscape scale regarding the water demand and irrigation water requirements in response to climate change and different urban farming location. Some projects (e.g. URBAN GreenUP; NativeScapeGR, Nature4-Cities, CITYFOOD) focus on mitigation the effects and risks of climate change and improving the water management cities. Other projects contribute to a better understanding of ecosystem services through evaluation and mapping the urban environments (TUNESinURB, Nedkov *et al.* 2017).

A better understanding how irrigation is used in the urban farming is needed to reduce pressures on limited freshwater resources, based on the knowledge how to use it efficiently based on economics, yield, environmental and social issues, aesthetics and safety for human health criteria. Assessment of underlying spatial and temporal variability in water use decisions at a landscape scale regarding the water demand and irrigation water requirements in response to climate change and different urban farming location.

What appears to be missing in most projects are concepts and experiences of water storage, rainwater harvesting and optimised usage of water with state-of-the-art irrigation systems. Very few long-term experiences exist at the moment for combining urban agriculture and any wastewater usage – as would fit within the concept of the circular city. As many current projects are using hydroponic systems, more research into a combination of hydroponic systems and wastewater reuse could be beneficial.

Air pollution

The economic growth, industrialisation and urbanisation has caused increased concentrations of pollutants such as ozone (O_3), nitrous oxides (NOx), sulphur dioxide (SO_2) and suspended particulate matter (SPM) in urban areas. These gases can cause significant damage to crops.

Suspended particulate matter has the most significant effect on crop yields and the quality of the crops. Several studies ascertained the correlation between the atmospheric pollutant deposition originating from traffic emissions (e.g. Cd, Cu, Cr, Ni, Pb and Zn) and the concentration of heavy metals in the vegetation and soil samples. Therefore it is necessary to assess the suitability of produce grown near pollution sources for human consumption and also identify causal links between a specific pollutant and damage to crops.

Education and knowledge transfer

Combined circular food systems – aquaculture and plant production together in the same system – are entirely new in the perspective of food production techniques and can also be implemented widely in education and knowledge transfer (Junge *et al.* 2019). In Norway, researchers have focused on the development of recirculating aquaculture systems (RAS) during the past 30 years, and the production of the most salmonids fingerlings grow in land-based RAS. In NIBIO Landvik, an aquaponics facility is operated with salmonids (brown *Salmo trutta*, rainbow trout *Oncorhynchus mykiss* and a relict salmon species called 'bleke' *Salmo salar* (Barlaup 2011), together with a wide range of plant species, such as wild herbs, Asian greens, edible flowers, leafy plants and several lettuce varieties (Skar *et al.* 2015). The system is used for research and education, and works as a showcase for an innovative approach to a more sustainable food production. The latest development is to apply the system to prisons, to create jobs, build skills and build prisoners' social acceptance in city communities by producing local and healthy food for the surrounding community (Skar 2018). Together with the system in Norway, further aquaponic systems were constructed for educational and knowledge transfer purposes, e.g. within the CITYFOOD project in Brazil and Germany and within the AQU@TEACH project in the United Kingdom, Slovenia, Spain and Switzerland. These so-called 'living labs' are serving as perceptible demonstration sites and are a central part of a diverse communication and dissemination agenda.

Knowledge gaps and further research question in urban agriculture

Consolidating the current knowledge on urban agriculture in green urban systems is needed. More knowledge is needed on multifunctionality and the relation to green infrastructure and food-productive urban landscapes, circular city debates and discussions of the possible adverse effect of air pollution on urban agriculture products' quality. Improved sustainability in the cities by integrating with buildings and waste conversion sites.

To encourage more holistic solutions to the problem, we present a comprehensive overview of European projects and examples which consider circular city approaches on urban farming. Examples of solutions from our expertise and database COST Action is found here: http://www.circular-city.eu/images/pdf_download/Proceedings_COST_WS_13-15Feb.pdf.

Due to the mainstreamed globalisation of urban food systems, the entire production and marketing schemes have shifted to comfort the needs of globalised value chains. Changes like these in the built environment and the socio-cultural practice create path dependencies. A significant gap of research on what a change in food production, processing, distribution and consumption can and will do to the cities, especially in the area of transportation/logistics, green infrastructure, resource streams and all the physical requirements on which they are based (streets, shops, pipes, wires, channels and more). We also identified a gap in food governance research and realised an almost complete absence of the social dimension meaning the analysis of socio-cultural patterns and practices in the food preparation and consumption.

doi: 10.2166/bgs.2019.931

ACKNOWLEDGEMENTS

The work is funded within the COST Action CA17133 Circular City ('Implementing nature-based solutions for creating a resourceful circular city', http://www.circular-city.eu, duration 22 Oct 2018–21 Oct 2022). COST Actions are funded within the EU Horizon 2020 Programme. The authors are grateful for the support and very early contribution of this article: T. Schvartz, Z. Heuschel, A. Helm, E. Paton, H. Paulenz, L. Straigyte and A. Canet-Martí.

SUPPLEMENTARY MATERIAL

The Supplementary Material for this paper is available at http://dx.doi.org/10.2166/bgs.2019.931.

REFERENCES

Agha Rokh, A. 2008 Evaluation of ornamental flowers and fishes breeding in Bushehr urban wastewater using a pilot-scale aquaponic system. *Water and Wastewater* **19** (65), 47–53.

Agrawal, M., Singh, B., Rajput, M., Marshall, F. & Bell, J. N. B. 2003 Effect of air pollution on peri-urban agriculture: a case study. *Environmental Pollution* **126** (3), 323–329. https://www.sciencedirect.com/science/article/pii/S0269749103002458#aep-section-id24.

Ainsworth, A. E. 2008 Rice production in a changing climate: a meta-analysis of responses to elevated carbon dioxide and elevated ozone concentration. *Globe Change Biology*. https://onlinelibrary.wiley.com/doi/pdf/10.1111/j.1365-2486.2008.01594.x<10.9.2019.

Aubry, C. & Kebir, L. 2013 Shortening food supply chains: a means for maintaining agriculture close to urban areas? The case of the French metropolitan area of Paris. *Journal of Food Policy* **41**, 85–93.

Aubry, C., Kebir, L. & Pasquier, C. 2012 *Le raccourcissement des circuits alimentaires: une nouvelle ruralité en périphérie des villes?* (étude de cas en île-de-France). In: *Nouveaux Rapports à la Nature Dans Lescampagnes* (Papy, F., Mathieu, N. & Férault, C. eds). Editions Quae, Paris, pp. 55–69.

Bachman, G. R. & Metzger, J. D. 2008 Growth of bedding plants in commercial potting substrate amended with vermicompost. *Bioresource Technology* **99** (8), 3155–3161.

Bailey, A., Williams, N., Palmer, M. & Geering, R. 2000 The farmer as a service provider: the demand for agricultural commodities and equine services. *Journal of Agricultural Systems* **66**, 191–204. doi:10.1016/S0308-521X(00)00047-0.

Barbieri, C. & Mahoney, E. 2009 Why is diversification an attractive farm adjustment strategy? Insights from Texas farmers and ranchers. *Journal of Rural Studies* **25**, 58–66. doi: 10.1016/j.jrurstud.2008.06.001.

Barlaup, B. T. 2011 *Byglandsbleke Salmo Salar*. Fact sheet: Artsdatabankens faktaark no. 175. http://www2.artsdatabanken.no/faktaark/Faktaark175.pdf.

Barrett, G. E., Alexander, P. D., Robinson, J. S. & Bragg, N. C. 2016 Achieving environmentally sustainable growing media for soilless plant cultivation systems – A review. *Scientia Horticulturae* **212**, 220–234.

Beauchesne, A. & Bryant, C. R. 1999 Organic farming: combining economic, social and environmental objectives in metropolitan regions. In: *Progress in Research on Sustainable Rural Systems* (Bowler, I., Bryant, C. R. & Firmino, A. eds). Série Estudos, No. 2, Universidade Nova de Lisboa, Centro de Estudos de Geografia e Planeamento Regional. https://www.researchgate.net/publication/268872462_Organic_farming_combining_economic_social_and_environmental_objectives_in_metropolitan_regions/stats.

Bohn, K. 2016 *PHVision: Development Vision for Patrick-Henry-Village Heidelberg*. University of Brighton. https://iba.heidelberg.de/de/projekte/landwirtschaftspark.

Boons, F. & Lüdeke-Freund, F. 2013 Business models for sustainable innovation: state-of-the-art and steps towards a research agenda. *Journal of Cleaner Production* **45**, 9–19.

Brown, S. L., Chaney, R. L. & Hettiarachchi, G. M. 2016 Lead in urban soils: a real or perceived concern for urban agriculture? *Journal of Environmental Quality* **45** (1), 26–36.

Bruun, E. W., Ambus, P., Egsgaard, H. & Hauggaard-Nielsen, H. 2012 Effects of slow and fast pyrolysis biochar on soil C and N turnover dynamics. *Soil Biology and Biochemistry* **46**, 73–79.

Buehler, D. & Junge, R. 2016 Global trends and current status of commercial urban rooftop farming. *Sustainability* **8** (11), 1108.

Chauhan, A. & Joshi, C. P. 2010 Effect of ambient air pollutants on wheat and mustard crops growing in the vicinity of an urban and industrial area. *New York Science Journal* **3** (2), 52–60. https://pdfs.semanticscholar.org/0af3/d227dda90d61f4c7e9cc62b1d1f5383cf20c.pdf.

Chen, Z., Ngo, H. H. & Guo, W. 2013 A critical review on the end use of recycled water. *Journal of Environmental Science and Technology* **43** (14), 1446–1516. doi:10.1080/10643389.2011.647788.

Clarke, L. W., Jenerette, G. D. & Bain, D. J. 2015 Urban legacies and soil management affect the concentration and speciation of trace metals in Los Angeles community garden soils. *Environmental Pollution* **197**, 1–12. doi: 10.1016/j.envpol.2014.11.015.

Corcoran, M. P. & Calvin, J. S. 2015 Introduction. In: *Urban Agriculture Europe* (Lohrberg, F., Licka, L., Scazzosi, L. & Timpe, A. eds). Jovis, Berlin, pp. 56–57.

Daigger, G. 2009 Evolving urban water and residual management paradigms: water reclamation and reuse, decentralization, resource recovery. *Water Environment Research* **81** (8), 809–823.

Deelstra, T. & Girardet, H. 2000 Urban agriculture and sustainable cities. In: *Growing Cities, Growing Food: Urban Agriculture on the Policy Agenda. A Reader on Urban Agriculture* (Bakker, N., Dubbeling, M., Gündel, S., Sabel-Koschella, U. & de Zeeuw, H. eds). Deutsche Stiftung für Internationale Entwicklung (DSE), Zentralstelle für Ernährung und Landwirtschaft, pp. 43–65. Available from: https://www.researchgate.net/publication/284992045_Urban_agriculture_and_sustainable_cities.

De Kimpe, C. R. & Morel, J. L. 2000 Urban soil management: a growing concern. *Journal of Soil Science* **165** (1), 31–40.

Delaide, B., Delhaye, G., Dermience, M., Gott, J., Soyeurt, H. & Jijakli, M. H. 2017 Plant and fish production performance, nutrient mass balances, energy and water use of the PAFF Box, a small-scale aquaponic system. *Aquacultural Engineering* **78**, 130–139.

Delgado, C. 2018 Contrasting practices and perceptions of urban agriculture in Portugal. *International Journal of Urban Sustainable Development* **10**, 170–185. doi: 10.1080/19463138.2018.1481069.

Despommier, D. 2010 *The Vertical Farm: Feeding the World in the 21st Century*. Picador, New York.

Drechsel, P., Keraita, B., Cofie, O. O. & Nikiema, J. 2015 Productive and safe use of urban organic wastes and wastewater in urban food production systems in low-income countries. In: *Cities and Agriculture – Developing Resilient Urban Food Systems* (de Zeeuw, H. & Drechsel, P. eds). RUAF Foundation, International Water Management Institute, Leusden, The Netherlands, pp. 162–191. Available from: https://www.ruaf.org/sites/default/files/7.%20Productive%20and%20safe%20use-min.pdf.

Ecosystem Services – TEEB 2015, available at http://www.teebweb.org/resources/ecosystem-services/ (accessed 20 May 2015).

Elgåker, H. & Wilton, B. 2008 Horse farms as a factor for development and innovation in the urban-rural fringe with examples from Europe and Northern America. 10th annual conference. In: *Nordic-Scottish University for Rural and Regional Development*, 2007 03 08–2007 03 10, Vejle, Denmark. Innovation systems and rural development, pp. 43–55.

Ericksen, P. J. 2008 Conceptualizing food systems for global environmental change research. *Global Environmental Change* **18** (1), 234–245. doi:10.1016/j.gloenvcha.2007.09.002.

Eriksen-Hamel, N. & Danso, G. 2010 Agronomic considerations forurban agriculture in southern cities. *International Journal of Agricultural Sustainability* **8** (1–2), 86–93. doi:10.3763/ijas.2009.0452. https://www.tandfonline.com/doi/pdf/10.3763/ijas.2009.0452.

Escala, M., Zumbuhl, T., Koller, C., Junge, R. & Krebs, R. 2012 Hydrothermal carbonization as an energy-efficient alternative to established drying technologies for sewage sludge: a feasibility study on a laboratory scale. *Energy & Fuels* **27** (1), 454–460.

European Commission 2010 *A Strategy for Smart, Sustainable and Inclusive Growth*. No. COM(2010) 2020 Final. https://eur-lex.europa.eu/LexUriServ/LexUriServ.do?uri=COM:2010:2020:FIN:EN:PDF.

European Commission 2011 *The EU Biodiversity Strategy to 2020, Environment*. Publ. Off. of the Europ. Union, Luxembourg. http://ec.europa.eu/environment/nature/info/pubs/docs/brochures/2020%20Biod%20brochure%20final%20lowres.pdf.

European Commission 2013a *Green Infrastructure (GI) - Enhancing Europe's Natural Capital*, Communication from the Commission No. COM(2013) 249 final. http://ec.europa.eu/environment/nature/ecosystems/docs/green_infrastructures/1_EN_ACT_part1_v5.pdf.

European Commission 2013b *Technical Information on Green Infrastructure (GI)*. Commission Staff Working Document No. SWD(2013) 155 final. http://ec.europa.eu/environment/nature/ecosystems/docs/green_infrastructures/1_EN_autre_document_travail_service_part1_v2.pdf.

European Commission 2019 *Implementation of the Circular Economy Action Plan*. Report from the commission to the European parliament, the council, the European economic and social committee and the committee of the regions. https://ec.europa.eu/environment/circular-economy/pdf/report_implementation_circular_economy_action_plan.pdf.

FAO 2019 *Urban Agriculture: FAO's Role in Urban Agriculture*. Available online: fao.org/urban-agriculture/en/ (accessed on 10.09.2019).

Fletcher, T. D., Andrieu, H. & Hamel, P. 2013 Understanding, management and modelling of urban hydrology and its consequences for receiving waters: a state of the art. *Advances in Water Resources* **51**, 261–279. doi: 10.1016/j.advwatres.2012.09.001.

Game, I. & Primus, R. 2015 *Global Sustainable Development*. Report, 2015 Edition. Advance Unedited Version. Economic and Social Affairs, United Nations, p. 198.

Gardner, B. L. 1994 Commercial agriculture in metropolitan areas: economics and regulatory issues. *Cambridge Journal of Agricultural and Resource Economics Review* **23** (1), 100–109. doi: https://doi.org/10.1017/S1068280500000460.

Gold, M., Cunningham, M., Bleuler, M., Arnheiter, R., Schönborn, A., Niwagaba, C. & Strande, L. 2018 Operating parameters for three resource recovery options from slow-pyrolysis of faecal sludge. *Journal of Water, Sanitation and Hygiene for Development* **8** (4), 707–717. https://doi.org/10.2166/washdev.2018.009.

Golden, H. E. & Hoghooghi, N. 2018 Green infrastructure and its catchment-scale effects: an emerging science. *WIRES Water* **5** (1), 14. doi: 10.1002/wat2.1254.

Graber, A. & Junge, R. 2009 Aquaponic systems: nutrient recycling from fish wastewater by vegetable production. *Desalination* **246** (1–3), 147–156. doi:10.1016/j.desal.2008.03.048.

Gruda, N. & Schnitzler, W. H. 2006 Wood fibres as a peat alternative substrate for vegetable production. *Eur. J. Wood Prod.* **64** (5), 347–350. https://doi.org/10.1007/s00107-005-0067-7.

Hamel, E. N. & Danso, G. 2011 Agronomic considerations for urban agriculture in southern cities. *International Journal of Agricultural Sustainability* **8**, 1–2. Source: https://doi.org/10.3763/ijas.2009.0452.

Harland, I., Lane, S. & Price, D. 1999 Further experiences with recycled zeolite as a substrate for the sweet pepper crop. *Acta Horticulturae* **481**, 187–194. doi:10.17660/ActaHortic.1999.481.19.

Heimlich, R. E. & Barnard, C. H. 1992 Agricultural adaptation to urbanization: farm types in northeast metropolitan areas. *Northeastern Journal of Agricultural and Resource Economics* **21** (1), 50–60.

Herrera-Gomez, S. S., Quevedo-Nolasco, A. & Pérez-Urrestarazu, L. 2017 The role of green roofs in climate change mitigation. A case study in Seville (Spain). *Building and Environment* **123**, 575–584. doi:10.1016/j.buildenv.2017.07.036.

Houston, P. 2005 Re-valuing the fringe: some findings on the value of agricultural production in Australia's peri-urban regions. *Journal of Geographical Research* **43** (2), 209–223. https://doi.org/10.1111/j.1745-5871.2005.00314.x.

Inwood, S. M. & Sharp, J. S. 2012 Farm persistence and adaptation at the rural-urban interface: succession and farm adjustment. *Journal of Rural Studies* **28**, 107–117.

IPBES 2019 *Summary for Policymakers of the Global Assessment Report on Biodiversity and Ecosystem Services of the Intergovernmental Science-Policy Platform on Biodiversity and Ecosystem Services.* https://www.ipbes.net/system/tdf/spm_global_unedited_advance.pdf?file=1&type=node&id=35245.

Jean-Soro, L., Le Guern, C., Béchet, B., Lebeau, T. & Ringeard, M.-F. 2015 Origin of trace elements in an urban garden in Nantes, France. *Journal of Soils and Sediments.* **15** (8), 1802–1812. doi:10.1007/s11368-014-0952-y.

Jensen, M. H. 2010 *Controlled Environment Agriculture in Deserts, Tropics, and Temperate Regions – A World Review.* Supported by CEAC, the Controlled Environment Agricultural Center, College of Agriculture and Life Sciences, and The University of Arizona, USA, p. 9.

Junge, R. & Graber, A. 2014 *The Potential of Aquaponics for Food Production in the Cities of the Future.* Publication, ZHAW/University of Ljubljana. https://pd.zhaw.ch/publikation/upload/207531.pdf.

Junge, R., Griessler Bulc, T., Anseeuw, D., Yavuzcan Yildiz, H. & Milliken, S. 2019 Aquaponics as an educational tool. In: *Aquaponics Food Production Systems: Combined Aquaculture and Hydroponic Production Technologies for the Future* (Goddek, S., Joyce, A., Kotzen, B. & Burnell, G. M. eds). Springer Nature Switzerland, pp. 561–595. https://link.springer.com/chapter/10.1007/978-3-030-15943-6_22.

Kennedy, G., Nantel, G. & Shetty, P. 2004 *Globalization of Food Systems in Developing Countries: A Synthesis of Country Case Studies.* Globalization of Food Systems in Developing Countries: Impact on Food Security and Nutrition. FAO Food and Nutrition Paper 83, pp. 1–26. Available from: http://www.fao.org/3/a-y5736e.pdf.

Khalil, S. & Kakar, M. K. 2011 Agricultural use of untreated urban wastewater in Pakistan. *Asian Journal of Agriculture and Rural Development* **1** (1), 21–26.

Khan, F., Khan, M. J., Samad, A., Noor, Y., Rashid, M. & Jan, B. 2015 In-situ stabilization of heavy metals in agricultural soils irrigated with untreated wastewater. *Journal of Geochemical Exploration* **159**, 1–7. doi:10.1016/j.gexplo.2015.07.002.

Krucker, M., Hummel, R. L. & Cogger, C. 2010 Chrysanthemum production in composted and non composted organic waste substrates fertilized with nitrogen at two rates using surface and sub-irrigation. *HortScience* **45** (11), 1695–1701. doi:10.21273/HORTSCI.45.11.1695.

Krupa, S., McGrath, M. T., Andersen, C. P., Booker, F. L., Burkey, K. O., Chappelka, A. H., Chevone, B. I., Pell, E. J. & Zilinskas, B. A. 2001 Ambient ozone and plant health. *Plant Disease* **85**, 4–12.

Lennard, W. A. & Leonard, B. V. 2006 A comparison of three different hydroponic sub-systems (gravel bed, floating and nutrient film technique) in an aquaponic test system. *Aquaculture International* **14** (6), 539–550.

Lwasa, S., Mugagga, F., Wahab, B., Simon, D., Connors, J. & Griffith, C. 2014 Urban and peri-urban agriculture and forestry: transcending poverty alleviation to climate change mitigation and adaptation. *Urban Climate* **7**, 92–106. doi:10.1016/j.uclim.2013.10.007.

Maher, M., Prasad, M. & Raviv, M. 2008 Organic soilless media components. In: *Soilless Culture: Theory and Practice* (Raviv, M. & Lieth, J. H. eds). Elsevier, Amsterdam, pp. 459–504.

Maucieri, C., Nicoletto, S., Sambo, Z., Komives, P., Borin M, T. & Junge, R. 2017 Vegetable intercropping in a small-scale aquaponic system. *Agronomy* **7** (4), 63. doi: 10.3390/agronomy7040063.

Maucieri, C., Forchino, A. A., Nicoletto, C., Junge, R., Pastres, R., Sambo, P. & Borin, M. 2018a Life cycle assessment of an aquaponic system built using recovered material for learning purposes. *Journal of Cleaner Production* **172**, 3119–3127. doi.org/10.1016/j.jclepro.2017.11.097.

Maucieri, C., Nicoletto, C., Junge, R., Schmautz, Z., Sambo, P. & Borin, M. 2018b Hydroponic systems and water management in aquaponics: a review. *Italian Journal of Agronomy* **13**, 1012. Available from: https://www.agronomy.it/index.php/agro/article/view/1012, https://doi.org/10.4081/ija.2017.1012.

Maucieri, C., Nicoletto, C., van Os, E., Anseeuw, D., Van Havermaet, R. & Junge, R. 2019 *Hydroponic Technologies.* In: *Aquaponics Food Production Systems: Combined Aquaculture and Hydroponic Production Technologies for the Future* (Goddek, S., Joyce, A., Kotzen, B. & Burnell, G. M. eds). Springer Nature Switzerland AG, pp. 77–110. https://doi.org/10.1007/978-3-030-15943-6_4.

McClintock, N. 2010 Why farm the city? Theorizing urban agriculture through a lens of metabolic rift. *Cambridge Journal of Regions Economy and Society* **3** (2), 191–207. doi: 10.1093/cjres/rsq005.

McEldowney, J. 2017 *Urban Agriculture in Europe: Patterns, Challenges and Policies. In-Depth Analysis.* Report European Parliament, EPRS (European Parliamentary Research Service), Brussels. PE 614.641. doi:10.2861/413185.

McPhearson, T., Andersson, E., Elmqvist, T. & Frantzeskaki, N. 2015 Resilience of and through urban ecosystem services. *Ecosystem Services* **12**, 152–156. doi:10.1016/j.ecoser.2014.07.012.

Michel, J. C. 2010 The physical properties of peat: a key factor for modern growing media. *Mires and Peat* **6**. HAL Id: Hal-00729716. https://hal-agrocampus-ouest.archives-ouvertes.fr/hal-00729716.

Million, A., Bürgow, G. & Steglich, A. 2016 Urban waters for urban agriculture – ROOF WATER-FARMs as participatory and multifunctional infrastructure. In: *Sustainable Urban Agriculture and Food Planning* (Roggema, R. ed.). Routledge, New York, pp. 142–165.

Ming, D. W. & Mumpton, F. A. 1989 Zeolites in soils. In: *Minerals in Soil Environments* (2nd edn, Dixon, J. B. & Weed, S. B. eds). SSSA Book Series, no. 1. Soil Science Society of America, pp. 73–911. https://dl.sciencesocieties.org/publications/books/abstracts/sssabookseries/mineralsinsoile/873/preview/pdf.

Monsees, H., Kloas, W. & Wuertz, S. 2017 Decoupled systems on trial: eliminating bottlenecks to improve aquaponic processes. *PloS ONE* **12** (9), e0183056. https://doi.org/10.1371/journal.pone.0183056.

Monsees, H., Suhl, J., Paul, M., Kloas, W., Dannehl, D. & Würtz, S. 2019 Lettuce (*Lactuca sativa*, variety Salanova) production in decoupled aquaponic systems: same yield and similar quality as in conventional hydroponic systems but drastically reduced greenhouse gas emissions by saving inorganic fertilizer. *PLoS ONE* **14** (6), e0218368. https://doi.org/10.1371/journal.pone.0218368.

Moragues-Faus, A., Sonnino, R. & Marsden, T. 2017 Exploring European food system vulnerabilities: towards integrated food security governance. *Environmental Science & Policy* **75**, 184–215. doi:10.1016/j.envsci.2017.05.015.

Mougeot, L. J. A. 2000 *Urban Agriculture: Definition, Presence, Potentials and Risks, and Policy Challenges.* Cities Feeding People Series, Report 31, p. 62. https://idl-bnc-idrc.dspacedirect.org/bitstream/handle/10625/26429/117785.pdf?sequence=12.

Mougeot, L. J. A. 2001 Urban agriculture: Definitions, presence, potentials and risks. In: *Growing Cities, Growing Food, Thematic Paper 1: Urban Agriculture on the Policy Agenda: A Reader on Urban Agriculture* (Bakker, N. *et al.* eds). German Foundation for International Development (DSE), Feldafing, Germany, pp. 1–42. https://www.ruaf.org/sites/default/files/Theme1_1_1.PDF.

Mumenthaler, C. 2015 *Developing A Methodology to Analyse and Compare Governance of Urban Agriculture.* STSM report COST Action Urban Agriculture Europe, p. 23. http://www.urban-agriculture-europe.org/files/stsm_report_mumenthaler.pdf (accessed on 15/08/2019).

Murray, A. & Buckley, C. 2010 Designing reuse-oriented sanitation infrastructure: the design for service planning approach. In: *Wastewater Irrigation and Health – Assessing and Mitigating Risks in low-Income Countries* (Drechsel, P., Scott, C. A., Raschid-Sally, L., Redwood, M. & Bahri, A. eds). Earthscan, International Water Management Institute (IWMI), International Development Research Centre (IDRC), pp. 303–318. http://publications.iwmi.org/pdf/H042615.pdf.

Nabulo, G., Oryem-Origa, H. & Diamond, M. 2006 Assessment of lead, cadmium and zinc contamination of roadside soils, surface films, and vegetables in Kampala City, Uganda. *Environmental Research* **101**, 42–52. doi: 10.1016/j.envres.2005.12.016.

Nair, R., Varghese, S. H., Nair, B. G., Maekawa, T., Yoshida, Y. & Kumar, D. S. 2010 Nanoparticulate material delivery to plants. *Plant Sci* **179** (3), 154–163. doi: 10.1016/j.plantsci.2010.04.012.

Ndiokwere, C. L. 1984 A study of heavy metal pollution from motor vehicle emissions and its effect on roadside soil, vegetation and crops in Nigeria. *Environmental Pollution Series B, Chemical and Physical* **7** (1), 35–42. https://doi.org/10.1016/0143-148X(84)90035-1.

Nedkov, S., Zhiyanski, M., Dimitrov, S., Borisova, B., Popov, A., Ihtimanski, I., Yaneva, R., Nikolov, P. & Bratanova-Doncheva, S. 2017 Mapping and assessment of urban ecosystem condition and services using integrated index of spatial structure. *Research Article in One Ecosystem* **2**, e14499. doi: 10.3897/oneeco.2.e14499.

Némethy, S., Bittsánszky, A., Schmautz, Z., Junge, R. & Kőmíves, T. 2016 Protecting plants from pests and diseases in aquaponic systems. In: *Ecological Footprint in Central Europe. Multiauthorial Monograph* (Łabaj, M. & Sobczuk, J. eds). The University College of Tourism and Ecology Press, Sucha Beskidzka, Poland, pp. 166–179. Available from: https://pbn.nauka.gov.pl/sedno-webapp/works/780004#.

NiChualain, D., Carlile, W., Hynes, C., Phelan, G., O'Haire, R. & Doyle, O. P. E. 2011 Nutrient status of co-composted indigenous Irish wastes, and their use in growing media. *Acta Horticulturae* **891**, 85–92. doi:10.17660/ActaHortic.2011.891.8.

Norton-Brandão, D., Scherrenberg, S. M. & van Lier, J. B. 2013 Reclamation of used urban waters for irrigation purposes – A review of treatment technologies. *Journal of Environmental Management* **122** (15), 85–98. https://doi.org/10.1016/j.jenvman.2013.03.012.

Nozzi, V., Graber, A., Schmautz, Z., Mathis, A. & Junge, R. 2018 Nutrient management in aquaponics: comparison of three approaches for cultivating lettuce, mint and mushroom herb. *Agronomy* **8** (3), 27. https://doi.org/10.3390/agronomy8030027.

Okorogbona, A. O. M., Denner, F. D. N., Managa, L. R., Khosa, T. B., Maduwa, K., Adebola, P. O., Okorogbona, A. O. M., Denner, F. D. N., Managa, R. L., Khosa, B. T., Maduwa, K., Adebola, P. O., Amoo, S. O., Ngobeni, H. M. & Macevele, S.

2018 Water quality impacts on agricultural productivity and environment. In: *Sustainable Agriculture Reviews 27* (Lichtfouse, E. ed.). Springer, Cham, pp. 1–35.

Ortolo, M. 2017 *Air Pollution Risk Assessment on Urban Agriculture*. Master thesis, WUR, NL. https://www.wur.nl/upload_mm/7/c/2/f753d12f-e73d-4a88-b4ac-8988b661283f_MSc_thesis_Ortolo%20%28final%29.pdf.

Oteng-Peprah, M., Acheampong, M. A. & deVries, N. K. 2018 Greywater characteristics, treatment systems, reuse strategies and user perception – a review. *Water Air & Soil Pollution* **229**, 255. doi:10.1007/s11270-018-3909-8.

Paço, T. A., Cruz de Carvalho, R., Arsénio, P. & Martins, D. 2019 Green roof design techniques to improve water use under Mediterranean conditions. *Urban Sci.* **3**, 14. doi:10.3390/urbansci3010014.

Perez-Murcia, M. D., Moral, R., Moreno-Caselles, J., Perez-Espinosa, A. & Paredes, C. 2006 Use of composted sewage sludge in growth media for broccoli. *Bioresource Technology* **97** (1), 123–130. https://doi.org/10.1016/j.biortech.2005.02.005.

Pérez-Urrestarazu, L., Fernández-Cañero, R., Franco-Salas, A. & Egea, G. 2015 Vertical greening systems and sustainable cities. *Journal of Urban Technology* **22** (4), 65–85. doi:10.1080/10630732.2015.1073900.

Piorr, A., Zasada, I., Doernberg, A., Zoll, F. & Ramme, W. 2018 *Research for AGRI Committee: Urban and Peri-Urban Agriculture in the EU – A Study Requested by the AGRI Committee*. Policy Department for Structural and Cohesion Policies, Directorate General for Internal Policies of the Union, Brussels. PE 617.468. doi:10.2861/953859.

Pölling, B., Alfranca, Ó., Alves, E., Andersson, G., Branduini, P., Egloff, L. D., Giacchè, G., Heller, A., Herkströter, K., Kemper, D., Koleva, G., Lorleberg, W., Mendes-Moreira, P., Miguel, A., Neves, L., Paulen, O., Pickard, D., Prados, M. J., Recasens, X., Ronchi, B., Spornberger, A., Timpe, A. G., Torquati, B., van der Schans, J. W., Weissinger, H. & Wydler, H. 2015 Creating Added Value Societal Benefits of Urban Agriculture. In: *Urban Agriculture Europe* (Lohrberg, F., Licka, L., Scazzosi, L. & Timpe, A. eds). Jovis, Berlin, pp. 92–100.

Pothukuchi, K. & Kaufman, J. L. 1999 Placing the food system on the urban agenda: The role of municipal institutions in food systems planning. In: *Agriculture and Human Values*, Vol. 16. Kluwer Academic Publishers, Printed in the Netherlands, pp. 213–224. doi:10.1023/A:1007558805953.

Prain, G. & de Zeeuw, H. 2007 Enhancing technical, organisational and institutional innovation in urban agriculture. *Urban Agriculture Magazine* **19**, 9–15.

Pratt, T., Allen, L. N., Rosenberg, D. E., Keller, A. A. & Kopp, K. 2019 Urban agriculture and small farm water use: case studies and trends from Cache Valley, Utah. *Agricultural Water Management* **213** (C), 24–35. doi:10.1016/j.agwat.2018.09.034.

Provė, C., Kemper, D., Loudiyi, S., Mumenthaler, C. & Nikolaidou, S. 2015 Governance of Urban Agriculture Initiatives: Insights drawn from European case studies. In: *Urban Agriculture Europe* (Lohrberg, F., Licka, L., Scazzosi, L. & Timpe, A. eds). Jovis, Berlin, pp. 64–69.

Qadir, M., Wichelns, D., Raschid-Sally, L., Singh Minhas, P., Drechsel, P., Bahri, A. & McCornick, P. 2007 Agricultural use of marginal-quality water – opportunities and challenges. In: *Water for Food Water for Life – A Comprehensive Assessment of Water Management in Agriculture* (Molden, D. ed.). Earthscan, International Water Management Institute (IWMI), pp. 425–457. Available from: https://www.researchgate.net/publication/237526608_Agricultural_use_of_marginal-quality_water-_opportunities_and_challenges.

Rakocy, J. E. 2012 Aquaponics – Integrating Fish and Plant Culture. In: *Aquaculture Production Systems* (Tidwell, J. ed.). John Wiley & Sons, Inc., pp. 343–386. Available from: https://onlinelibrary.wiley.com/doi/abs/10.1002/9781118250105.ch14.

Raviv, M. 2013 Composts in growing media: what's new and what's next? *Acta Horticulturae* **982**, 39–47.

Rich, S. 1964 Ozone damage to plants. *Annual Review of Phytopathology* **2**, 253–266.

Rogers, R. 1997 *Cities for A Small Planet*. Faber & Faber Limited, London, p. 180

Roggema, R. 2016 Sustainable Urban Agriculture and Food Planning, Routledge Studies in Food. In: *Society and the Environment* (Roggema, R. ed.). Taylor and Francis, Milton, pp. 286. Available from: https://doi.org/10.4324/9781315646039.

Rosenzweig, C., Solecki, W., Hammer, S. A. & Mehrotra, S. 2010 Cities lead the way in climate–change action. *Nature – International Journal of Science* **467** (7318), 909–911. https://www.nature.com/articles/467909a.pdf.

Samson, R., Ningal, T. F., Tiwary, A., Grote, R., Fares, S., Saaroni, H., Hiemstra, J. A., Zhiyanski, M., Vilhar, U., Cariñanos, P., Järvi, L., Przybysz, A., Moretti, M. & Zürcher, N. 2017 Species-Specific Information for Enhancing Ecosystem Services. In: *The Urban Forest, Future City 7* (Pearlmutter, D. *et al.* eds). Springer International Publishing, pp. 111–144. doi:10.1007/978-3-319-50280-9_12.

Santo, R., Palmer, A. & Brent, K. 2016 *Vacant Lots to Vibrant Plots: A Review of the Benefits and Limitations of Urban Agriculture*. Baltimore. doi:10.13140/RG.2.2.25283.91682.

Sanyé-Mengual, E., Anguelovski, I., Oliver-Solà, J., Montero, J. I. & Rieradevall, J. 2015a Resolving differing stakeholder perceptions of urban rooftop farming in Mediterranean cities: promoting food production as a driver for innovative forms of urban agriculture. *Agriculture and Human Values*. 3–22. doi: 10.1007/s10460-015-9594-y.

Savvas, D. & Passam, H. 2002 *Hydroponic Production of Vegetables and Ornamentals*. Embryo publications, Athens, Greece, p. 463.

Schmautz, Z., Loeu, F., Liebisch, F., Graber, A., Mathis, A., Griessler Bulc, T. & Junge, R. 2016 Tomato productivity and quality in aquaponics: comparison of three hydroponic methods. *Water* **8** (11), 533. https://doi.org/10.3390/w8110533.

Schmilewski, G. 2008 The role of peat in assuring the quality of growing media. *Mires & Peat* **3**, 1–8. Available from: https://pdfs.semanticscholar.org/bcdf/aa9ba59ef6c39c52a4cb1e81962f2cf33171.pdf.

Schwarz, K., Pickett, S. T. A., Lathrop, R. G., Weathers, K. C., Pouyat, R. V. & Cadenasso, M. L. 2012 The effects of the urban built environment on the spatial distribution of lead in residential soils. *Environmental Pollution* **163**, 32–39.

Simon-Rojo, M., Recasens, X., Callau, S., Duží, B., Eiter, S., Hernández Jiménez, V., Kettle, P., Laviscio, R., Lohrberg, F., Pickard, D., Scazzosi, L. & Cavallo, A. 2015 From Urban Food Gardening to Urban Farming. In: *Urban Agriculture Europe* (Lohrberg, F., Licka, L., Scazzosi, L. & Timpe, A. eds). Jovis Verlag GmbH, Berlin, Germany, pp. 22–28.

Skar, S. L. G. 2018 *CITYFOOD – Smart Integrated Multi-Trophic City Food Production Systems –Water- and Energy-Saving Approach for Global Urbanization.* Project (Norwegian part). https://cityfood.igb-berlin.de/.

Skar, S. L. G., Liltved, H., Kledal, P. R., Høgberget, R., Björnsdottir, R., Oddsson, S., Homme, J. M., Drengstig, A., Savidov, N., Seljåsen, R. & Paulsen, H. 2015 *Aquaponics NOMA (Nordic Marin) - New Innovations for Sustainable Aquaculture in the Nordic Countries.* Project no. 11090, Nordic Innovation publication, 2015/06. Report 108 pages. http://norden.diva-portal.org/smash/get/diva2:1296888/FULLTEXT01.pdf.

Smit, J. 1996 Cities that feed themselves. In: *Urban Agriculture, Food, Jobs and Sustainable Cities* (2nd edn 2001, Smit, J., Ratta, A. & Nasr, J. eds). United Nations Development Program (UNDP) Publication, New York, pp. 1–29.

Soriano, A., Pallarés, S., Pardo, F., Vicente, A. B., Sanfeliu, T. & Bech, J. 2012 Deposition of heavy metals from particulate settleable matter in soils of an industrialised area. *Journal of Geochemical Exploration* **113**, 36–44. https://doi.org/10.1016/j.gexplo.2011.03.006.

Specht, K., Siebert, R. & Thomaier, S. 2015a Perception and acceptance of agricultural production in and on urban buildings (ZFarming): a qualitative study from Berlin, Germany. *Agriculture and Human Values.* doi:10.1007/s10460-015-9658-z.

Specht, K., Siebert, R., Thomaier, S., Freisinger, U. B., Sawicka, M., Dierich, A., Henckel, D. & Busse, M. 2015b Zero-acreage farming in the city of Berlin: an aggregated stakeholder perspective on potential benefits and challenges. *Sustainability* **7**, 4511–4523. doi:10.3390/su7044511.

Tahir, S., Steichen, T. & Martin, S. 2018 *Water and Circular Economy: A White Paper* (Jeffries, N. ed.). https://nextgenwater.eu/wp-content/uploads/2018/10/Water_and_circular_economy-Co.Project_White_paper.pdf.

Timpe, A., Cieszewska, A., Supuka, J. & Tóth, A. 2015 Urban Agriculture Goes Green Infrastructure. In: *Urban Agriculture Europe* (Lohrberg, F., Licka, L., Scazzosi, L. & Timpe, A. eds). Jovis Verlag GmbH, Berlin, pp. 126–137.

Torabian, S., Zahedi, M. & Khoshgoftar, A. H. 2017 Effects of foliar spray of nano-particles of $FeSO_4$ on the growth and ion content of sunflower under saline condition. *Journal of Plant Nutrition* **40** (5), 615–623. doi:10.1080/01904167.2016.1240187.

Tóth, A. & Timpe, A. 2017 Exploring urban agriculture as a component of multifunctional green infrastructure. Application of figure-ground plans as a spatial analysis tool. *Moravian Geographical Reports* **25** (3). doi:10.1515/mgr-2017-0018.

UN Habitat 2012 *Urban Patterns for A Green Economy: Working with Nature.* UN Habitat, Online, Nairobi. http://mirror.unhabitat.org/pmss/listItemDetails.aspx?publicationID=3341.

United Nations 2018 *Worlds Urbanizations Prospects: The 2018 Revisions.* Key Factors. Population Division of the UN Department of Economic and Social Affairs. https://population.un.org/wup/Publications/Files/WUP2018-Methodology.pdf.

UNU-IAS United Nations University Institute for Advanced Studies 2010 *Cities, Biodiversity and Governance: Perspectives and Challenges of the Implementation of the Convention on Biological Diversity at the City Level, Policy Report,* UN-UIAS, Yokohama. https://www.researchgate.net/publication/259679414_Cities_Biodiversity_and_Governance_Perspectives_and_Challenges_of_the_Implementation_of_the_Convention_of_Biological_Diversity_at_the_City_Level.

van der Schans, J. W., Lorleberg, W., Alfranca, O., Alves, E., Andersson, G., Branduini, P., Egloff, L., Giacchè, G., Heller, H., Herkströter, K. & Timpe, A. 2015 It Is a Business! Business Models in Urban Agriculture. In: *Urban Agriculture Europe* (Lohrberg, F., Licka, L., Scazzosi, L. & Timpe, A. eds). Jovis Verlag GmbH, Germany, pp. 82–91.

van Huylenbroeck, G., van Hecke, E., Meert, H., Vandermeulen, V., Verspecht, A., Vernimmen, T., Boulanger, A. & Luyten, S. 2005 Development strategies for a multifunctional agriculture in peri-urban areas. In: *Agriculture in an Urbanizing Society, vol 1. Proceedings of the Sixth AESOP Conference on Sustainable Food Planning* (Roggema, R. ed.), Cambridge Scholars Publishing, 695 pp. Available from: https://www.belspo.be/belspo/organisation/Publ/pub_ostc/CPagr/rappCP18r_en.pdf.

van Leeuwen, E., Nijkamp, P. & de Noronha Vaz, T. 2010 The multifunctional use of urban greenspace. *Journal of Agricultural Sustainability* **8**, 20–25.

van Veenhuizen, R. & Danso, G. 2007 *Profitability and Sustainability of Urban and Peri-Urban Agriculture.* FAO Agricultural Management, Marketing and Finance Occasional Paper No 19. FAO, Rome.

Vejre, H., Eiter, S., Hernández-Jiménez, V., Lohrberg, F., Loupa-Ramos, I., Recasens, X., Pickard, D., Scazzosi, L. & Simon-Rojo, M. 2015 Can Agriculture Be Urban? In: *Urban Agriculture Europe* (Lohrberg, F., Licka, L., Scazzosi, L. & Timpe, A. eds). Jovis Verlag GmbH, Germany, pp. 18–21.

Verdonck, O. D., De Vleeschauwer, D. & De Boodt, M. 1981 The influence of the substrate to plant growth. In symposium on substrates in horticulture other than soils *in situ. Acta Horticulturae* **126**, 251–258.

Viljoen, A., Bohn, K. & Howe, J. 2005 *Continous Productive Urban Landscape: Designing Urban Agriculture for Sustainable Cities.* Architectural Press, Oxford.

Villarroel, M., Rodriguez Alvariño, J. M. & Duran Altisent, J. M. 2011 Aquaponics: integrating fish feeding rates and ion waste production for strawberry hydroponics. *The Spanish Journal of Agricultural Research* **9** (2), 537–545.

Vlachokostas, C., Nastisb, S. A., Achillasa, C., Kalogeropoulosa, K., Karmirisc, I., Moussiopoulosa, N., Chourdakisa, E., Baniasa, G. & Limperia, N. 2010 Economic damages of ozone air pollution to crops using combined air quality and GIS modelling. *Atmospheric Enviroment* **44** (28), 3352–3361. https://www.sciencedirect.com/science/article/pii/S135223101000484X.

Voulvoulis, N. 2018 Water reuse from a circular economy perspective and potential risks from an unregulated approach. *Current Opinion in Environmental Science & Health* **2**, 32–45. doi:10.1016/j.coesh.2018.01.005.

doi: 10.2166/bgs.2019.931

Wästfelt, A. & Zhang, Q. 2016 Reclaiming localisation for revitalising agriculture: a case study of peri-urban agricultural change in Gothenburg, Sweden. *Journal of Rural Studies* **47** (Part A), 172–185. doi: 10.1016/j.jrurstud.2016.07.013.

Werkenthin, M., Kluge, B. & Wessolek, G. 2014 Metals in European roadside soils and soil solution – A review. *Environmental Pollution* **189**, 98–110. https://doi.org/10.1016/j.envpol.2014.02.025.

Yi, L., Jiao, W., Chen, X. & Chen, W. 2011 An overview of reclaimed water reuse in China. *Journal of Environmental Science* **23** (10), 1585–1593. https://doi.org/10.1016/S1001-0742(10)60627-4.

Zanin, G., Ponchia, G. & Sambo, P. 2009 Yield and quality of vegetables grown in a floating system for ready-to-eat produce. *Acta Horticulturae* **807**, 433–438. doi:10.17660/ActaHortic.2009.807.61.

Zanin, G., Evans, M. R., Bassan, A. & Sambo, P. 2012 Use of fresh rice hulls and anaerobic digestion residues as substrates alternative to peat. *Acta Horticulturae* **927**, 1003–1010.

Zasada, I. 2011 Multifunctional peri-urban agriculture – A review of societal demands and the provision of goods and services by farming. *Journal of Land Use Policy* **28** (4), 639–648. doi:10.1016/j.landusepol.2011.01.008.

Zheng, Y. & Chen, S. 1991 Atmospheric Environment (China) (in Chinese), Vol. 6, p. 45.

First received 30 June 2019; accepted in revised form 16 October 2019. Available online 7 November 2019

doi: 10.2166/bgs.2020.929

Transformation tools enabling the implementation of nature-based solutions for creating a resourceful circular city

Evina Katsou[a,*], Chrysanthi-Elisabeth Nika[a], Devi Buehler[b], Bruno Marić[c], Boldizsár Megyesi[d], Eric Mino[e], Javier Babí Almenar[f], Bilge Bas[g], Dženan Bećirović[c], Sabina Bokal[h], Maja Đolić[i], Nilay Elginöz[j], Gregoris Kalnis[k], Mari-Carmen Garcia Mateo[l], Maria Milousi[m], Alireza Mousavi[n], Iva Rinčić[o], Agatino Rizzo[p,q], Ignasi Rodriguez-Roda[r,s], Benedetto Rugani[f], Alvija Šalaševičienė[t], Ramazan Sari[u], Peyo Stanchev[a], Emel Topuz[v] and Natasa Atanasova[w]

[a] Department of Civil and Environmental Engineering, Institute of Environment, Health and Societies, Brunel University London, Uxbridge Campus, Middlesex, UB8 3PH, Uxbridge, UK

[b] Institute of Natural Resource Sciences (IUNR), ZHAW Zurich University of Applied Sciences, Grueental, 8820 Waedenswil, Switzerland

[c] University of Sarajevo, Faculty of Forestry, Zagrebačka 20, 71 000 Sarajevo, Bosnia and Herzegovina

[d] Centre for Social Sciences, Hungarian Academy of Sciences, TóthKálmán u. 4. 1095, Budapest, Hungary

[e] Technical Unit of the Euro-Mediterranean Information System on Know-how in the Water Sector, 06901 Sophia Antipolis, France

[f] Luxembourg Institute of Science and Technology (LIST) | Environmental Research & Innovation (ERIN) department | RDI Unit on Environmental Sustainability Assessment and Circularity (SUSTAIN) 41, Rue du Brill, L-4422 Belvaux, Luxembourg

[g] Department of Civil Engineering, Istanbul BilgiUniversity, Santralistanbul Campus, 34060 Istanbul, Turkey

[h] Global Water Partnership Central and Eastern Europe, Jeseniova 17, 833 15 Bratislava, Slovakia

[i] Faculty of Technology and Metallurgy, University of Belgrade, Karnegijeva 4, 10000 Belgrade, Serbia

[j] Department of Chemical Engineering, KTH Royal Institute of Technology, Brinellvägen 8, 114 28 Stockholm, Sweden

[k] Department of Architecture, School of Engineering, Frederick University, Nicosia, Cyprus

[l] MCG Research & Innovation Sustainability Architecture/Urban Planning, Valencia, Spain

[m] Department of Chemical Engineering, University of Western Macedonia, 50100, Kozani, Greece

[n] Department of Electronic and Computer Engineering, Institute of Engineering and Technology, Brunel University London, Uxbridge Campus, Middlesex, UB8 3PH, Uxbridge, UK

[o] University of Rijeka, Faculty of Medicine, BraćeBranchetta 20, 51 000 Rijeka, Croatia

[p] Architecture, Lulea University of Technology, 97187, Lulea, Sweden

[q] School of Landscape Architecture, Norwegian University of Life Sciences, 1430 Ås, Norway

[r] Catalan Institute for Water Research (ICRA), EmiliGrahit 101, 17003 Girona, Spain

[s] LEQUiA, Laboratory of Chemical and Environmental Engineering, University of Girona, Campus Montilivi, 17071, Girona, Spain

[t] Food Institute of Kaunas University of Technology, Radvilenu av. 19 C, 44239 Kaunas, Lithuania

doi: 10.2166/bgs.2020.929

[u] Faculty of Economics and Administrative Sciences, Middle East Technical University, Ankara, Turkey

[v] Department of Environmental Engineering, Gebze Technical University, 41400 Kocaeli, Turkey

[w] Faculty of Civil and Geodetic Engineering, University of Ljubljana, 1000 Ljubljana, Slovenia

*Corresponding author. E-mail: evina.katsou@brunel.ac.uk

Abstract

The linear pattern of production-consumption-disposal of cities around the world will continue to increase the emission of pollutants and stocks of waste, as well as to impact on the irreversible deterioration of non-renewable stocks of raw materials. A transition towards a circular pattern proposed by the concept of 'Circular Cities' is gaining momentum. As part of this urban transition, the emergent use of Nature-based Solutions (NBS) intends to shift public opinion and utilize technology to mitigate the urban environmental impact. In this paper, an analysis of the current research and practical investments for implementing NBS under the umbrella of Circular Cities is conducted. A combined appraisal of the latest literature and a survey of ongoing and completed National-European research and development projects provides an overview of the current enabling tools, methodologies, and initiatives for public engagement. It also identifies and describes the links between facilitators and barriers with respect to existing policies and regulations, public awareness and engagement, and scientific and technological instruments. The paper concludes introducing the most promising methods, physical and digital technologies that may lead the way to Sustainable Circular Cities. The results of this research provide useful insight for citizens, scientists, practitioners, investors, policy makers, and strategists to channel efforts on switching from a linear to a circular thinking for the future of cities.

Key words: assessment methods, Circular Cities, Nature-based Solutions, policies and regulations, stakeholders' awareness and engagement

ABBREVIATIONS

NBS	Nature-based Solutions
CE	Circular Economy
GI	Green Infrastructure
ICT	Information and Communication Technology
DSS	Decision Support Systems
KESI	Key Environmental and Socio-Economic Indicators
LCA	Life Cycle Assessment
LCC	Life Cycle Costing
MFA	Material Flow Analysis
CBA	Cost Benefit Analysis
PD	Participatory Design
UM	Urban Metabolism
HHA	Harvest to Harvest Approach
SS	Self-Sufficiency
MAES	Mapping and Assessment of Ecosystems and their Services
S-LCA	Social Life Cycle Assessment
ES	Ecosystem Services
IO	Input-Output
NW	Networking
RIA	Research and Innovation Action
IA	Innovation Action
CSA	Coordination and Support Action
FET	Future and Emerging Technologies
II	Implementation and Integration
TRLs	Technology Readiness Levels
WG	Working Group

SDGs Sustainable Development Goals
R&I Research and Innovation

INTRODUCTION

Cities are complex systems under continuous evolution, whose internal dynamics and process interactions generate impacts on the population's health, socio-economic well-being, and the environment (Alberti 2008; Pickett *et al.* 2008). Nowadays these impacts are mainly driven by a 'linear' behaviour according to which exploitable resources are transported to the cities to provide the necessities of urban consumption and waste dynamics. The patterns of acquisition, consumption and the subsequent waste disposal of resources pose environmental and socio-economic implications. These patterns may also cause irreversible deteriorations of non-renewable stocks of raw materials and huge waste discharged in the outflow of the urban system (Brunner & Rechberger 2016). The concept of 'circularity' is, therefore, gaining popularity among urban planners and decision makers (Petit-Boix & Leipold 2018; Prendeville *et al.* 2018; Williams 2019; Zeller *et al.* 2019) to counter the imbalances caused by unsustainable linear practices. The effect of consuming diminishing raw materials at a faster rate than the ability of nature to restore is a matter of serious global concern.

Concurrently, the emergent concept of Nature-based Solutions (NBS) promotes the circular use of resources making use of closed nutrient, water, and energy cycles by reusing waste rather than discarding it (EC 2015a). In urban areas, NBS could reinforce economic growth, which is highly dependent on the quantity and quality of natural resources, as well as on their availability (González-Val & Pueyo 2019), by promoting the sustainable use of natural resources and by harnessing natural processes (Connor 2015). The enhanced natural capital, as well as the efficient use of resources (i.e. energy and materials), facilitated by working with nature, would further build on the circular economy (CE) (UN 2018). Since CE initiatives at urban scale aim at transforming cities into sustainable and circular systems (Petit-Boix & Leipold 2018), NBS following the concept of CE as an intermediate link, can be seen as enablers to the transition from linear to Circular Cities. In this paper, we refer to NBS as defined in Langergraber *et al.* (2020).

Traditionally, cities have been shaped by institutions (local, regional or national) to comply with regulations, while fostering socio-economic development. Current urban policies, legislation and regulations are generally written in and for a linear economy thus, they may (unintentionally) hinder the transition to a CE. According to Stewart *et al.* (2016), policies and regulations can hamper CE by providing: (i) unclear or fuzzy messages, (ii) a complex system of changing regulations (e.g. multiple sectoral and interacting regulations on water, energy, waste, environment impact assessment), (iii) low pressure and a lack of control, and (iv) a limited space for innovation. Additionally, NBS are still poorly addressed by current policies and regulations related to CE. This may be due to the fact that NBS is a relatively novel concept in dealing with the challenges faced by society (i.e. societal challenges), and therefore, still searching for its place under different policies and regulations. NBS aims to address challenges associated with climate resilience, health and well-being in urban areas (IUCN 2012; Cohen-Schacham *et al.* 2016), integrating established ecosystem-based approaches, such as biodiversity, ecosystem services (ES), green infrastructure (GI) etc., and aiming in broadening them, in order to holistically tackle issues of environmental, economic and social nature in building resilience (Raymond *et al.* 2017a, 2017b). Within European Research and Innovation (R&I) programmes, e.g. 'Horizon 2020', NBS extend the aforementioned approaches involving biodiversity and ecosystem services aligned with goals of innovation for growth and job creation and work towards sustainable societal development (Maes & Jacobs 2017; Nesshöver *et al.* 2017). Therefore, due to NBS systemic nature there are inherent difficulties in integrating such a complex concept in policies and regulations (Niță *et al.* 2017). Unsurprisingly,

doi: 10.2166/bgs.2020.929

supporting frameworks for NBS are provided by other instruments, such as the EU green infrastructure strategy and the biodiversity strategy.

However, policies and legal frameworks can also accelerate the transition towards Circular Cities supporting solutions, such as NBS. Such an approach appears at the EU level since the Paris agreement with a set of EU strategies and policy papers connecting sectoral regulations and initiating a revision of existing Directives when further coherency is necessary. Thus, the EU Circular Economy package, launched in 2015 with an action plan composed of 54 items, is presented as a tool to foster the transition and encompass most of the barriers. Two key instruments are foreseen by the EC report on the implementation of the Circular Economy Action Plan (COM 2019). (i) Investments in innovation and in adapting the industrial base; with R&I (H2020), Environment (LIFE) programmes, Cohesion Policy or financing facilities. To overcome the regulation limits, pilot innovation deals have been introduced in 2016 and should be extended. (ii) Strong stakeholder engagement to ensure co-design and social acceptance of solutions.

In this regard, stakeholders play a central role in supporting the transition from linear to circular pathways in cities. The long-term viability and durability, the extent of scalability for the adoption and implementation of actions, projects, and/or solutions, as well as the diffusion of good practices are perceived as key in achieving and maintaining this transition (Kabisch *et al.* 2016). The involvement and collaboration of public, private and civil actors in the governance of NBS – enabling Circular Cities – can reduce barriers to NBS adoption on a wider scale of application (Frantzeskaki *et al.* 2014; Kabisch *et al.* 2016). The partnering of different actors can moreover improve the circulation of knowledge regarding NBS and Circular Cities (Ugolini *et al.* 2015). Citizens' participation and involvement can further facilitate the communication of information on NBS and their diffusion by the community (Kabisch *et al.* 2016). In this context, the use of participatory evaluation can be seen as a way to respect the legitimacy of different views on NBS quality, as well as to apply multiple perspectives provided by the different stakeholders (Nesshöver *et al.* 2017).

Another issue related to the relatively new concepts of NBS and Circular Cities is identified in the persisting challenges of the costs and benefits of circularity methods in cities (Raymond *et al.* 2017a, 2017b). Although the costs of the shift are tangible and measurable, the burden of proof on the short-term benefits of NBS is still on the proponents. Assuming that the long-term (e.g. resilience and well-being) impact is understandable by stakeholders, the intangible short-term benefits are more difficult to convey. It seems that one of the major shortcomings is the lack of holistic and widely accepted methodology and/or framework to assess the circularity potential of such systems into the bargain, as argued by (Kabisch *et al.* 2016). There is a need for reaching consensus on the individual instruments and tools (methods, indicators, models, databases, etc.) capable of comparing linearity against circularity. The aim is to offer a roadmap and the tools to converge methodologies and integrated functionalities to estimate the costs and benefits of circular against existing linear solutions. The roadmap paves the way and the tools to provide the means to achieve the goal. The proposed toolbox would contain methods and models to quantify a set of Key Environmental and Socio-Economic Indicators (KESI) integrated by Information and Communication Technology (ICT) and in the form of software applications to aid in decision making (i.e. Decision Support Systems). These platforms will become the instruments to compare circular against linear scenarios, and demonstrate short-term and long-term tangible and intangible benefits of modern Circular Cities. Such objective measurements will become the drivers of public awareness/education and investment.

The goal of this review paper – developed within the framework of the COST Action Circular City (CA17133 2018) – is threefold: (i) to identify and group specific tools and methods that are used to assess NBS for implementing and improving circularity in cities, (ii) to identify means of society and stakeholders' engagement and awareness, and (iii) to identify barriers and facilitators within current policies and regulations in order to promote and enable the implementation of NBS for improving future city transitions. The identification of current gaps, needs and opportunities will help to transfer research results into the market and to upscale existing pilot applications into the ground of concrete decision making for NBS.

doi: 10.2166/bgs.2020.929

METHODOLOGY

To identify the current gaps and opportunities regarding the transformation tools enabling the implementation of NBS for creating Circular Cities as stated in the introduction, an analysis was performed that includes: (i) the interconnections between policies and regulations, engagement and participation of stakeholders, and tools and assessment methodologies, during the implementation of NBS (see section The Scope of Nature-based Solutions in Circular Cities); (ii) a literature review focusing on the up-to-date scientific research (see section Overview); and (iii) a survey of projects related to NBS for creating Circular Cities that would enable the comparison between scientific research and practical applications (see section Survey Results and Discussion). The paper focuses on the survey of National-European research and development projects, which sets a delimitation of this study with regards to NBS contributions to sustainable research and development in European context.

Literature review approach

The conducted literature review (see section Overview) is divided into three sub-sections, namely Policy and Regulations, Stakeholder Engagement and Awareness, and Tools and Methods. The review is based on peer-reviewed papers published in international scientific journals and on high-level policy documents. The identification of the relevant papers was conducted using related key words in the Science Direct and Scopus databases. High-level policy documents were chosen as the planning and implementation of NBS is supported by such documents.

The focus of the review is to: (i) identify current policies and regulations at EU level that should be conserved when implementing NBS for Circular Cities, as well as to understand the importance of the social, economic and environmental dimensions to be considered in and inform the policies (see section Policy and Regulations); (ii) identify the challenges related to public awareness and social acceptance, as well as to review methods that have been developed to increase the stakeholders engagement in NBS (see section Stakeholder Engagement and Awareness); and (iii) identify promising tools and methods that have been developed by researchers and can be used to assess the different dimensions (environmental, economic, and social) of the effectiveness of NBS (see section Tools and Methods).

Project survey approach

The members of this COST Action hold key knowledge on different aspects of NBS and Circular Cities and they participate in national, European and academic projects related to these subjects. Therefore, past and ongoing projects were reviewed and a meta-analysis of policies and regulations, stakeholders' awareness and participation, as well as the tools and methods assessing the technologies and systems was conducted.

Data from the projects were collected through a specially designed online questionnaire, which was carried out during March 2019. The main aim of the survey was to collect useful information on the different projects in order to compare and identify the gaps between scientific literature and practical applications.

The questions included in the questionnaire were divided in four main categories. The first category included questions related to general information of the projects, such as:

- Project title;
- Project type – the projects are divided by: (i) Networking (NW), (ii) Research and Innovation Action (RIA) and Innovation Action (IA), (iii) Coordination and Support Action (CSA), (iv) Future and Emerging Technologies (FET), and (v) Implementation and Integration (II);
- Project stage – from recently started to completed;

doi: 10.2166/bgs.2020.929

- NBS focus – four categories, corresponding to the working groups (WG) of this COST Action, were specified: Built Environment (WG1), Urban Water (WG2), Resource Recovery (WG3), and Urban Farming (WG4);
- Implementation level – i.e. conceptual, experimental, on-ground, and capitalization projects;
- Application scale – *micro* (technology, material, energy, etc.), *meso* (building, neighbourhood, landscape, etc.), and *macro* (district, city, region, etc.);
- TRL of the different technologies – the scale ranges from 1 to 9 with TRL 1: basic principles observed and reported, and TRL 9: system ready for full-scale deployment;
- SRL of the project – the scale ranges from 1 to 9 with SRL 1: identifying problem and identifying societal readiness, and SRL 9: actual project solution(s) proven in relevant environment.

The second category was related to the methods that were used for the assessment of the NBS and the ICT tools that were either developed or used in the projects. The third category was focused on the type of policies and regulations that were considered in the project, the main barriers that were identified in the projects, as well as supporting measures (of policy or regulations) to ensure the success of the project. Finally, the fourth category was related to the type of the stakeholders involved in the project, the barriers for the implementation of project activities related to stakeholder awareness and engagement, the types of public participation tools and techniques used in the project and the level of stakeholders' engagement in the project.

A total number of 47 relevant research projects were studied (the list of the reviewed projects can be found in Appendix 1) and the results of the conducted survey are presented under Survey Results and Discussion.

THE SCOPE OF NATURE-BASED SOLUTIONS IN CIRCULAR CITIES

There are four main steps for the implementation of NBS in creating Circular Cities: (i) planning, (ii) design, (iii) assessment and, (iv) communication of results. As illustrated in Figure 1, it is important to close the circular process starting a new plan based on the information provided in the communication of previous implementations. Tools are used in all of the four steps in order to enhance the dialogue among the different actors, as well as the engagement and participation of the stakeholders to induce sustainable changes and assess the environmental, economic, and social improvement of the cities, and finally to inform the policy making.

The **planning** phase requires information about the regulations that must be fulfilled and the city where the NBS will be implemented aiming at becoming a circular city (specific problems to be solved, citizens' awareness and perception about the problem, characteristics of the city including potential sites to implement the NBS, possible/alternative NBS to implement, and similar case studies and demo sites implemented around the world). The typical tools used for planning include Decision Support Systems, Multi-Criteria Decision Analysis, repositories of case studies (literature or

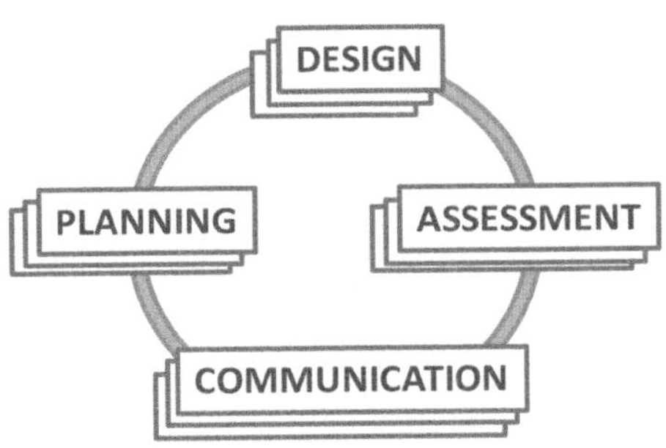

Figure 1 | Circular process for the implementation of NBS in Circular Cities.

web-based), models and databases with information about the former value of the indicators that will be used for the assessment, exploratory/visualization tools to determine the characteristics of the city/sites (e.g. aerial images, land space, green coverage, etc.), and participatory digital platforms. Most of these tools were applied in literature case studies, but they still have to overcome some serious challenges before reaching the market and being fully applied for decision makers (McIntosh *et al.* 2011; Poch *et al.* 2017).

The **design** and implementation requires the use of engineering design knowledge, process models and specific software. Involving general stakeholders and the specific neighbourhood citizens to co-design the NBS becomes crucial for the acceptance and engagement of the society.

The **assessment** phase includes process performance monitoring (sensors, instrumentation, automation, control) and any measurement or assessment of the impacts (beneficial or detrimental) of the implemented NBS (by using methodologies, such as Life Cycle Assessment – LCA, Life Cycle Costing – LCC, Cost Benefit Analysis – CBA, etc.). Moreover, specific software, web-based questionnaires and apps (e.g. citizen science, where the citizens become the 'sensors' for monitoring both the performance and the impact of the NBS) are usually used for this purpose.

Finally, the **communication** phase includes the dissemination to citizens and stakeholders (for their real engagement and in order to increase social perception about the new services of the Circular Cities) and any potential exploitation activities. Efficient communication should make the planning a living process, while revised planning based on communication starts the whole implementation cycle again. The most typical tools for communication are social media and web-based platforms that include repository databases, case studies, user-friendly Decision Support Systems (DSS), simulations, and participatory platforms. This step is the key to providing information to new cities that want to become resourceful and circular. Excellent science needs effective communication and dissemination. Bringing research and its outcomes to the attention of non-scientific audiences, scientific peers, potential business partners or policy makers fosters collaboration and innovation. Strategic communication and dissemination will help to explain the wider societal relevance of science, build support for future research and innovation funding, ensure uptake of results within the scientific community, and open up potential business opportunities for novel products or services.

OVERVIEW

Policy and regulations

Regulation and governance arrangements can be considered as tools supporting policy strategies. Figure 2 presents the different facets of these three pillars, i.e. policy, regulation and governance, which are necessary for the long-term stability of NBS for circular city initiatives.

The governance of NBS emerges as a complex phenomenon, involving multiple social and political actors, premises and visions. Defining the appropriate mix between the involvement of state, local authorities, private sector and grassroots movements remains challenging, in particular in terms of cost sharing and long-term sustainability. Participatory approaches with multiple stakeholders' impacts evaluation and a strong civil society engagement appears as a successful approach governance (Naturvation 2017).

In the context of this paper, a state-of-the-art analysis of EU regulations and policies is conducted. An inventory of a wide range of EU instruments to be conserved when implementing NBS for Circular Cities is presented in Table 1. The different EU regulations are categorized based on their content and they are linked relevantly to the four different challenges addressed by NBS, as presented in (Langergraber *et al.* 2020). Due to its wide diversity, the local dimension requires further research, in particular considering the relationships between Circular Cities and their surrounding

doi: 10.2166/bgs.2020.929

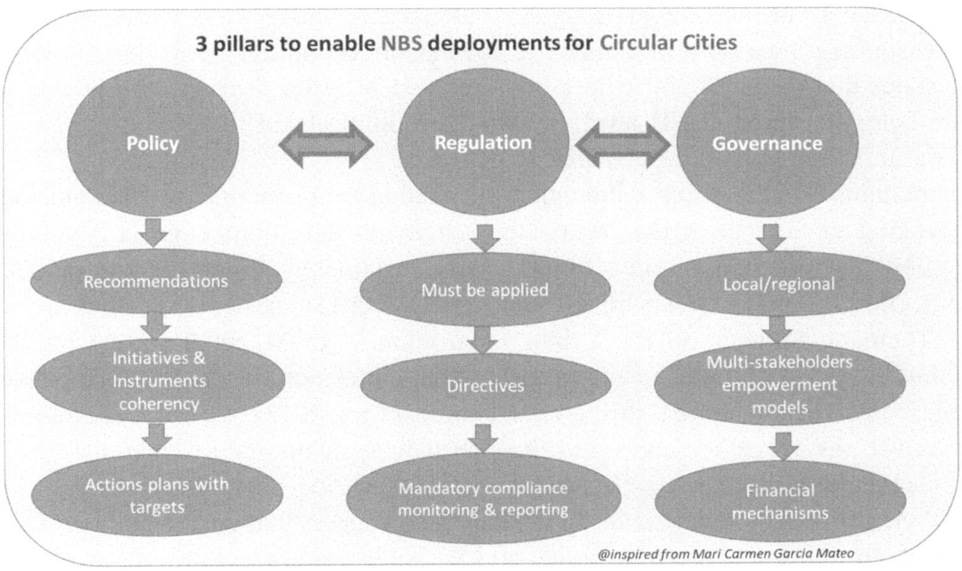

Figure 2 | Three pillars to enable NBS deployments for Circular Cities.

Table 1 | EU policies and regulations for the implementation of NBS in Circular Cities

EU policies / strategies	EU regulations	
• Circular Economy Package: Action Plan and Monitoring Framework • Green Infrastructure Strategy • Bio-economy strategy • Regional Development and Cohesion Policy • Biodiversity strategy • Common Agricultural Policy (CAP) • Framework Programme for Research and Innovation • Environment Action Programmes	Resource recovery	Waste Framework Directive (2008/98/EC)
	Built environment	Energy performance of buildings Directive (2010/31/EU) Energy efficiency Directive (2012/27/EU)
	Built environment Water Resource recovery Urban farming	Environmental Impact Assessment (EIA)
	Resource recovery Urban farming	Sewage Sludge Directive (86/278/EEC)
	Resource recovery Urban farming	Fertilisers Regulation (EC) No. 2003/2003
	Water Resource recovery Urban farming	Water Framework Directive (2000/60/EC) Groundwater Directive (2006/118/EC) Drinking Water Directive (98/83/EC), revised on 1 Feb 2018 Urban Waste Water Directive (91/271/EEC) Minimum requirements for water reuse COM (2018) 337 (2018/0169 (COD))

environment, such as the river basin when considering aquatic ecosystems or the whole region for secondary products created from urban circular economy (e.g. fertilizers derived from sludge or organic waste).

Regulations are often seen as barriers for innovation, but they are also providing an enabling environment for new product development and marketing.

UN and ISO non-binding frameworks support NBS implementation for Circular Cities. For example, the frameworks on 'treated wastewater for irrigation', 'drinking water quality guidelines' and 'Sustainable Development Goals' (SDGs) are worth mentioning. Often it is challenging translating such strategic and visionary frameworks at the operational level and in local contexts, especially in areas with weak planning and regulation systems (i.e. some areas in Central and Eastern Europe). Moreover, even the EU Directives contain many non-legally binding articles related to water reuse.

For example, Article 12 of the UWWD suggests that 'Treated wastewater shall be reused whenever appropriate. Disposal routes shall minimize the adverse effects on the environment'. However, it is the member states that decide if and where it is appropriate. After years of discussion, water reuse standards are being accepted at EU level, but they are only related to reuse in agriculture. Other uses (e.g. in urban areas) remain unregulated at EU level.

The EU regulations listed in Table 1 show that the need to protect natural capital and value ES are recognized as being crucial to progress towards the sustainable development goals. A notable example is represented by the European Union actions towards sustainable growth for Europe 2020 and EU Biodiversity (COM 2011) and Green Infrastructure (COM 2013) strategies. Furthermore, the European Union Thematic Strategy on the Urban Environment (COM 2005) recognizes that it is in urban areas that the environmental, economic and social dimensions of the EU Sustainable Development Strategy come together most strongly (Raymond *et al.* 2017a, 2017b). Therefore, NBS are directly relevant to several policy areas and through their systemic nature they interact with many others (land use, planning etc.).

The EU policies/strategies reported in Table 1 do also suggest that sustainability assessment has recently become an important issue for policy and decision makers in Europe due to a recognized requirement of balance between environmental, economic and social policies. The interlink between these policies requires a simultaneous consideration of the intersection between environment, economy and society of sustainable development to have a better environment, economic growth (ideally decoupled from resources exploitation) and welfare of society without compromising the wealth of future generations as indicated in Brundtland Commission definition of sustainable development (WCED Brundtland Commission 1987). There exists an inherent risk of new technology implementations on the balance of policies related to the three pillars of sustainability.

Although the social dimension to sustainability is widely recognized, performing a social assessment is difficult due to a lack of indicators that can be directly employed in technical analyses. Therefore, attention has been given to determination and quantification of social factors or the interaction of the social variables in a complex relationship. Without quantified and properly determined social factors, the impact of policies and technologies on the well-being of society and environment may not be a solid base for future policy strategies. Thus, the assessments of economic and environmental dimension without considering the social effects are insufficient (Carrera & Mack 2010). The social acceptance of technology, renewable energy and environmental policies are progressively becoming more important for policy and decision makers worldwide aiming to design policies that reach attempted targets smoothly with community support. Therefore, social acceptance could be considered to be a promising factor for social assessment. As an emerging solution to environmental problems, NBS related projects or technologies are subject to social acceptance. There is a need to consider aspects of urban management, governance, biodiversity etc. within a society and integrate diverse systems of knowledge and values for NBS design and implementation in order to be socially comprehensible and acceptable to a range of stakeholders (Maes & Jacobs 2017).

Stakeholder engagement and awareness

Public awareness and social acceptance of the NBS for Circular Cities is important for its proliferation and success in the future. Kabisch *et al.* (2016) identified that one of the major impediments on proliferation of NBS is the traditional structures of city departments and the 'sectoral language', which traps knowledge into 'sectoral silos'; or the so-called compartmentalization of professionals with different educational background and different objectives (Brink *et al.* 2016). Implementation of NBS requires cooperation across departments of the administration or between various actors with different, and sometimes, competing objectives. Therefore, there is an imperative need of an agreement on the societal values, based on which urban development will be planned and adopted.

In order to be able to communicate and create impact, all stakeholders should refer to universal: (a) definitions of the key concepts, (b) values and valuing system, (c) metrics and indicators, (d) benchmarks and points of reference. Such an approach creates a common language and the foundations of the information and knowledge to be shared and customized for stakeholders (e.g. public, investor, regulator, and policy maker).

Provided the message can be conveyed appropriately, the next challenge is to assess the willingness of stakeholders to accept and adopt the solutions offered to them. In other words, what kind of short- and long-term benefits against the investments are made. This is especially relevant when the former is relatively immediate and the latter takes longer to implement.

Wüstenhagen *et al.* (2007) suggest three elements of social acceptance: socio-political, community and market acceptance. Note that, the convergence of socio-political, community and economic factors determines the '*willingness to accept*'. Socio-political acceptance can be enabled through regulatory frameworks and government standards (Beck & Martinot 2004).The benchmarks and measurable indicators of socio-political factors should provide the framework for firstly examining the procedural justice, which refers to decision making with respect to the collective interests of all stakeholders (Walker 2009). The second indicator of socio-political factors of acceptance is the distributional justice. Social trust can be achieved by fair distribution of costs and benefits and equal rights of access to information by citizens and decision makers. Efforts are necessary to quantify socio-political acceptance, as this is one of the major components in order to achieve social acceptance (Rosso-Cerón & Kafarov 2015). In the case of NBS, even though a small but growing number of countries have adopted regulatory frameworks promoting them (WWAP 2018), in the majority of cases, a universal and precise legal framework for NBS and following procedures for stakeholder awareness and engagement is still lacking and thus, limiting the outputs of the projects.

Community acceptance stands for local stakeholders' impression of the benefits that new NBS technologies or circularity initiatives bring to their respective communities. Normally, the methods for an objective examination of community acceptance, is the level engagement with disseminated information about projects and technologies (e.g. relevant subscription on social media, specialist magazines, local media, etc.). Furthermore, active and voluntary engagement of the community with surveys, attending town hall meetings, focus groups are other modes of increasing awareness as well as gauging acceptance. Roddis *et al.* (2018), for example, focused on the community acceptance of onshore wind and solar farms implementation plans in Great Britain (for the years between 1990 and 2017) by composing a set of indicators.

Market acceptance mainly gauges consumers' utility towards paying or contributing to an initiative or product. Added to the consumers are the role of investors and the business-to-business relationships (e.g. value chains) and their perception of short- and long-term cost and benefit of resource allocation. Wüstenhagen *et al.* (2007) for example, provide an insight to the attitudes of international companies towards different initiatives.

As the engagement of stakeholders in NBS to realize Circular Cities is widely understood and increasingly highlighted in the literature, some methods that have been developed to increase the stakeholders' engagement have been additionally reviewed. Design thinking is one of them, which deploys the typical design workshop setting for iteratively prototyping ideas and can be inscribed within the umbrella of Participatory Design (PD). PD aims at incorporating end-users as full participants in development processes. Mazé (2007) compares PD to user-centered design, which draws on diverse means of studying, analyzing and incorporating user needs into product development. PD focuses on different means for bringing design processes, representations, and products to participation by stakeholders with diverse skills and expertise. In Scandinavia, Atelier (Binder *et al.* 2011) defines participatory design as an approach that attempts to involve end-users in the design process. The author characterizes DT similarly to Redström (2008) 'use before use'. Atelier's 'design things' is inspired by Schön's reflective practitioner (Schön 1983) following an iterative design process through envisioning, prototyping, and experiencing. Through these phases, participants undergo emotional and cognitive experiences and they express themselves by engaging in practical action together, in a group. The inclusion of creativity can take different forms in different participatory approaches used today (Rizzo *et al.* 2015).

According to a recent study on Food Policy Councils (Bassarab *et al.* 2019), the most widespread strategies to raise stakeholder engagement include encouraging community members to participate in actions realizing NBS, hosting public events and forums, surveying community members, distributing newsletters, developing specific community engagement strategies, or cross-promoting partner organizations' events. Time, lack of basic knowledge on NBS and lack of engagement plan are also decisive factors to achieve stakeholders' engagement on a high level.

Tools and methods

According to EC (EC 2015a), NBS can address one or multiple societal challenges in sustainable ways and simultaneously provide multiple co-benefits for health, the economy, society and the environment. Despite such strong belief, one can observe a severe lack of practical and targeted guidance for assessing the impacts of NBS within and across different societal challenges (Raymond *et al.* 2017a, 2017b, 2017c). Previous studies have either: (i) assessed the performance of NBS with regard to specific challenge areas, such as regulating urban surface runoff (Zölch *et al.* 2017); (ii) assessed the performance of NBS with regard to their multiple co-benefits and compared them to alternative solutions, e.g. the study of Liquete *et al.* (2016); (iii) examined a set of indicators that can be used to measure the effectiveness of NBS addressing a specific societal challenge and the co-benefits, e.g. the studies of Kabisch *et al.* (2016) and Xing *et al.* (2017); or (iv) developed conceptual frameworks that still lack operationalization (e.g. Raymond *et al.* 2017a, 2017b; Calliari *et al.* 2019).

The same issue arises for CE initiatives realizing Circular Cities (i.e. lack of practical and targeted guidance for assessing circularity initiatives), as the available published data is insufficient to assess these strategies at the city-scale. On the one hand, most of the cities' initiatives are accessible in the cities' web pages and databases, i.e. grey literature (e.g. Bastein *et al.* 2016; Glasgow Chamber of Commerce 2016; Mairie de Paris 2017; Sack-Nielsen 2018). On the other hand, not all of the previous studies, assessing the environmental performance of CE strategies – published on peer-reviewed journal articles – have framed their assessment within the CE context (Petit-Boix & Leipold 2018).

In this work, a transdisciplinary approach – since circularity and sustainability concepts necessarily depend upon the interaction between the three spheres of society, economy and environment – has been undertaken to identify the tools and methods that have been used to assess NBS enabling circularity transitions from the literature. After the first classification of these tools and methods from the literature, the additional projects' survey – presented under Overview – was used to identify the most common tools and methods that are actually used in projects. Therefore, this large set of instruments, potentially suitable to address many of the open transdisciplinary questions associated with the quantitative characterization of Circular Cities, was summarized into categories (as presented in Table 2).

The analysis of the tools and methods used in the projects indicates that there is an extensive diversity in the use of methods for the assessment of NBS; however, their application is mainly case specific without taking advantage of the complementary features that often those methods offer. This is the case, for example, for the majority of the methods emerging from the present survey, i.e. LCA and ES assessment with MAES. Their screening in the context of the selected projects suggests that more research efforts are needed to identify mutual strengths and integrate different modelling approaches, e.g. the cascade modelling framework (Potschin-Young *et al.* 2018), to address multiple challenges, such as the accounting for bundles of services at different spatial and temporal scales. Some recent literature focusing on the combination of LCA and ES methods, for example, fosters the alignment of existing ES

doi: 10.2166/bgs.2020.929

Table 2 | Description of the most common tools and methods for NBS assessment

Name of Tool/Method	Description from Literature	Scope of Assessment		
		Environmental	Social	Economic
Life Cycle Assessment (LCA)	'LCA addresses the environmental aspects and potential environmental impacts (e.g. use of resources and the environmental consequences of pollutant releases) throughout a product's life cycle from raw material acquisition through production, use, end-of-life treatment, recycling and final disposal (i.e. cradle-to-grave)' (ISO 14040 2006).	x		
Life Cycle Costing (LCC)	'LCC is a technique that assesses costs over the life cycle of a product or a system' (Rödger *et al.* 2018).			x
Cost Benefit Analysis (CBA)	'CBA compares the gains and losses associated with an investment project (a road, railway line, port, urban expansion, etc.) or with a policy, e.g. the setting of an environmental standard' (Pearce 1998).	x	x	x
Material Flow Analysis (MFA)	'MFA methodology evaluates the flow of materials entering and leaving a system and their impact in the environment' (Rincon *et al.* 2013).	x		x
Urban Metabolism (UM)	'Model that quantifies processes and allows the measurement of four main cycles or flows: Water, materials, energy and nutrients' (Kennedy *et al.* 2007; Pincetl *et al.* 2012).	x		
Harvest to Harvest Approach (HHA)	'Urban Harvesting reduces single source dependence by optimizing the demand and by harvesting local resources' (Wielemaker *et al.* 2018).	x		
Self-sufficiency (SS)	Self-sufficiency is achieved by reusing output as an input, (partially) covering the input demand. SSI can be used as a measure for the extent of self-sufficiency of a system (Wielemaker *et al.* 2018).	x		
Mapping and Assessment of Ecosystems and their Services (MAES)	'MAES is a conceptual framework that links socio-economic systems with ecosystems via the flow of ecosystem services and through the drivers of change that affect ecosystems either as consequence of using the services or as indirect impacts due to human activities in general' (Maes *et al.* 2013).	x	x	x
Social Life Cycle Assessment (S-LCA)	'A social and socio-economic Life Cycle Assessment (S-LCA) is a social impact (and potential impact) assessment technique that aims to assess the social and socio-economic aspects of products and their potential positive and negative impacts along their life cycle encompassing extraction and processing of raw materials; manufacturing; distribution; use; re-use; maintenance; recycling; and final disposal' (Andrews *et al.* 2009).		x	

classification systems to accepted life cycle inventories (LCIs), and the implementation of a consensual ES-LCA framework (Othoniel *et al.* 2016; Verones *et al.* 2017; Maia de Souza *et al.* 2018). At the same time, studies dealing with life cycle thinking combined with urban metabolism (UM) assessment and/or input-output (IO) analysis methods (e.g. Pincetl *et al.* 2012; Goldstein *et al.* 2013; Beloin-Saint-Pierre *et al.* 2017; Petit-Boix *et al.* 2017; Sohn *et al.* 2018) confirm the findings of the present review, i.e. that both UM and IO based models can arguably benefit from the accommodation of bottom-up LCI technology knowledge, which considers micro-scale details, although this is at the expense of increased modelling complexity. Other attempts to use the complementary features of methods is hybridization of UM and LCA. While MFA-based approaches, such as UM, HHA, SS, measure material flows to/out of a city or system, which is useful for the quantification of the circularity potential, it lacks assessing the environmental impacts of these flows. Thus, hybridization with LCA fills up this gap and it was implemented in Chester *et al.* (2012); Goldstein *et al.* (2013).

In this regard, the present analysis has further confirmed that it is worth adopting the main assets of LCA in the context of macro-scale assessments, e.g. the standardization criteria of the LCA method, the representativeness, transparency and completeness of the LCI databases, the necessary flexibility to host different types of flows in LCA models, not only physical but also monetary flows, etc. However, the isolated use of LCA and its family of methods (such as S-LCA, LCC, etc.) will always generate biases and a lack of consideration of the many landscape features, cultural diversity, and socio-economic attributes that characterize the cities as complex and dynamic systems. Therefore, a more top-down approach shall be undertaken to coupling LCA with other tools. As already highlighted by some works (e.g. Onat *et al.* 2017; Marvuglia *et al.* 2018; Beaussier *et al.* 2019), when combined with LCA tools, such as system dynamics, agent-based modelling, etc., can allow to capture the multifaceted features of those systems, bringing to more comprehensive studies of the urban metabolism as a showcase to establish circular city models.

Additionally to the previously described assessment methodologies, a number of decision support tools has been used aiming to facilitate the NBS implementation in cities. Most of them focus on sustainable urban drainage, i.e. on integrating water and green infrastructure to achieve multiple benefits. To name a few, Urban BEATS (Bach *et al.* 2015) simulates the planning, design and implementation of water sensitive urban design infrastructure in urban environments. In addition the Adaptation Support Tool (Voskamp & Van de Ven 2015) facilitates the collaborative planning towards more resilient and attractive environment. E2STORMED is a comprehensive decision support tool that applies Multi-Criteria Analysis (Morales-Torres *et al.* 2016) and includes a catalogue of more than 20 types of drainage infrastructures. Radinja *et al.* (2019) have introduced a DSS that supports design and evaluation of blue-green infrastructure based on hydrology-hydraulic modelling.

SURVEY RESULTS AND DISCUSSION

This section presents the results of the survey as described under Project Survey Approach. It is important to underline the fact that most of the information extracted from the target projects is self-reported, while the projects come from various scientific fields, scale of enforcement or lines of financing. Therefore, the survey results provide the basis for a broader analysis.

Figure 3 shows the general information about the surveyed projects, while results that specifically focus on Policy and Regulations, Stakeholder Engagement and Awareness, and Tools and Methods are presented in the following sub-sections.

Regarding the different project types – shown in Figure 3(a) – around 70% of the projects are self-identified as RIA and IA, indicating that there is a considerable number of pilot projects that are actually implemented. The majority of the projects have low to medium Technology Readiness Levels (TRLs) indicating that the maturity of either the investigated topic (NBS and related applications)

doi: 10.2166/bgs.2020.929

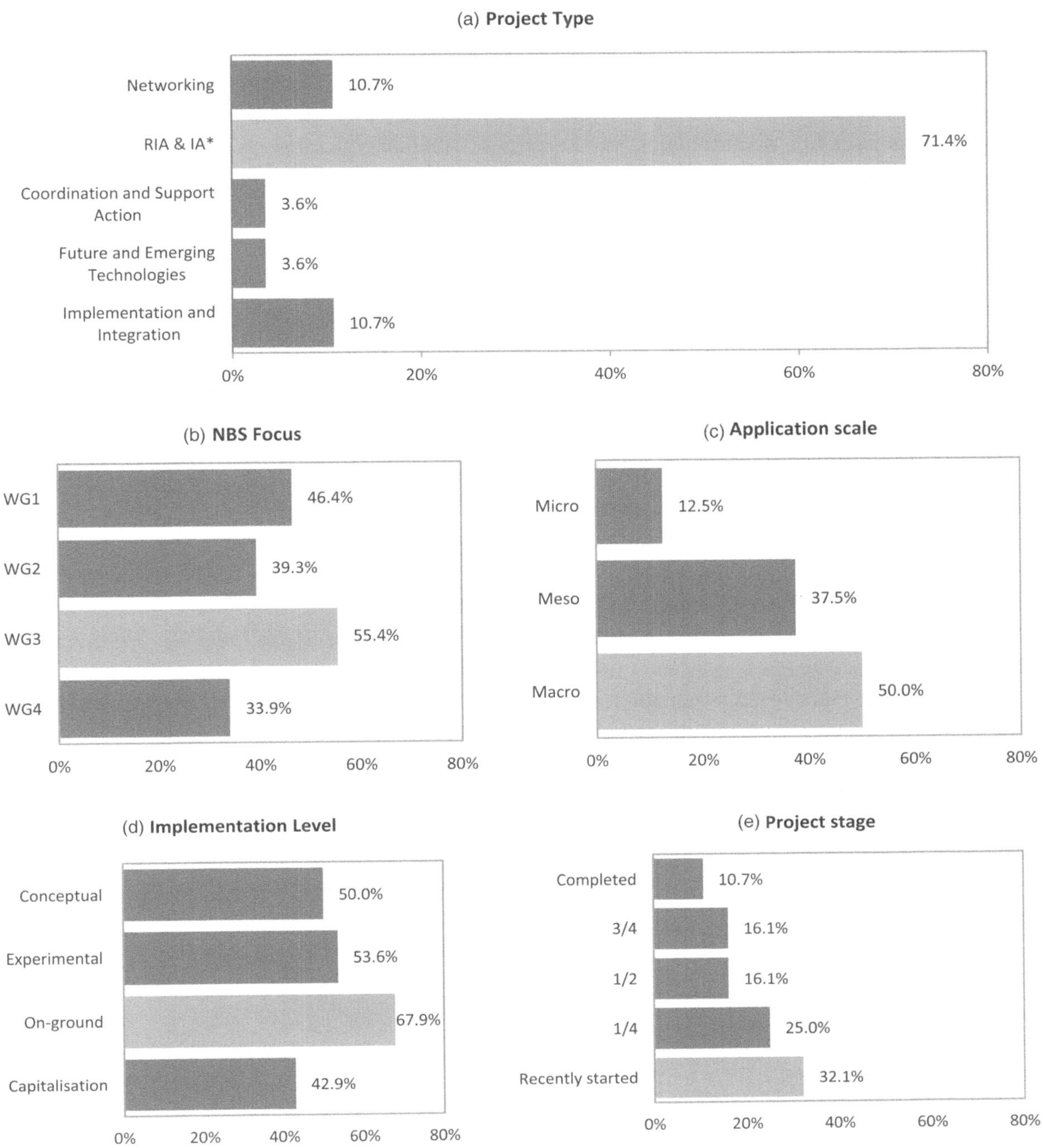

Figure 3 | Categorization of the selected projects. *RIA stands for Research and Innovation Action, IA stands for Innovation Action.

or the application of the method(s) still necessitates improvements and innovation. This may be explained by the fact that the concept of Circular Cities is quite recent, while the direct or indirect quantifiable, social/economical or ecological benefits captured by the implemented NBS projects are still quite unexplored (Marin & De Meulder 2018; Petit-Boix & Leipold 2018).

The second classification – related to the focus of NBS (Figure 3(b)) – results in an equal representation of all categories, which indicates that NBS for Circular Cities involve all sectors and disciplines in the urban development.

The results concerning the projects classification based on the scale of application (Figure 3(c)) indicate that half of the projects applied NBS at a city or regional scale, 37.5% of projects applied their technologies on a building or neighbourhood scale, while only 12.5% of the projects focused on the micro-scale.

The results regarding the implementation level (Figure 3(d)) demonstrate that different levels of implementation exist within the same project. Interestingly, a large number of projects (almost 68%) implemented their technologies on the ground (real engineering practice), which indicates the intention to provide empirical verification of the NBS concept.

Furthermore, the respondents stated the current project stage, i.e. level of completion ranging from just started to completed, at the time of the survey (Figure 3(e)). Out of the total of respondents, 10.7% of the projects are completed and 32.1% have started recently.

Policy and regulations

To understand the barriers or drivers that policies and regulations pose in practice, the following analysis was conducted. Among the 47 studied projects, 48% have considered European policies and regulations as the most important drivers or obligations for the project's implementation, followed by local regulations and governance (41%). Additionally, only 22% of the 47 projects undertook a policy/regulation review while, less than 10% have considered policy indicators for monitoring the success of their project.

Figure 4 shows the role of policy and regulation as drivers or barriers to NBS proliferation. Figure 4 further indicates that policy instruments, such as innovation, social, SDGs, and GI, are driving the changes while, more classical policies – linked to regulatory frameworks, such as water resources or environment – are considered as equally introducing drivers and barriers. It is worth noting that agriculture and biodiversity policies and regulations are perceived as more limiting rather than enabling the development of NBS. This can be due to NBS being developed in a close connection to the urban environment and its searches for sustainability or resilience, therefore NBS still need adjustments in spaces where the main objectives are different: biodiversity conservation or food production.

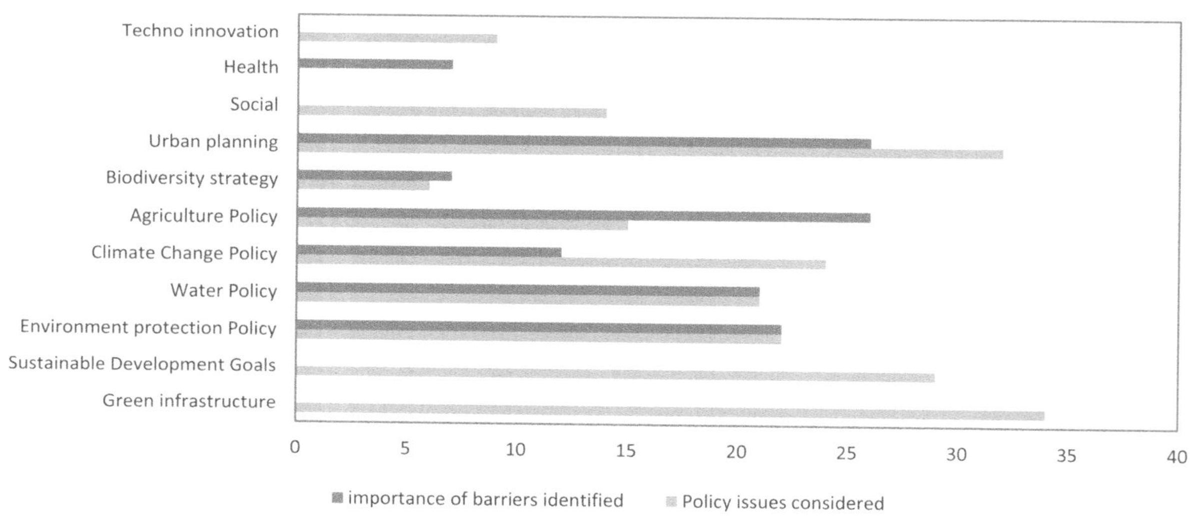

Figure 4 | Policies and regulations considered by projects and main barriers identified.

In addition to the survey conducted, Brink *et al.* (2016) identifies the lack of space in dense urban areas, environmental and building permits, and the possible conflict of interest with other ecosystem services (such as drinking water production), which can determine the appearance of ecosystem disservices in some situations (Schaubroeck 2018), as additional barriers to the NBS implementation.

doi: 10.2166/bgs.2020.929

Policies can include the creation of supporting framework, requirement or incentive to foster the development, implementation and deployment of NBS in cities. Figure 5 presents the types of such supporting measures that the projects surveyed are considering as positively impacting their development. Most of these projects are science driven, so unsurprisingly research and innovation frameworks are considered first, followed by political commitment and ownership. A suitable environment for market exploitation associated with financial incentives (grants and reduced taxes) is expected to boost NBS deployment in cities. Additionally, a long-term perspective of local governments on funding is necessary in order to create stability, decrease uncertainty for activities and enable voluntary action for a sustainable transition.

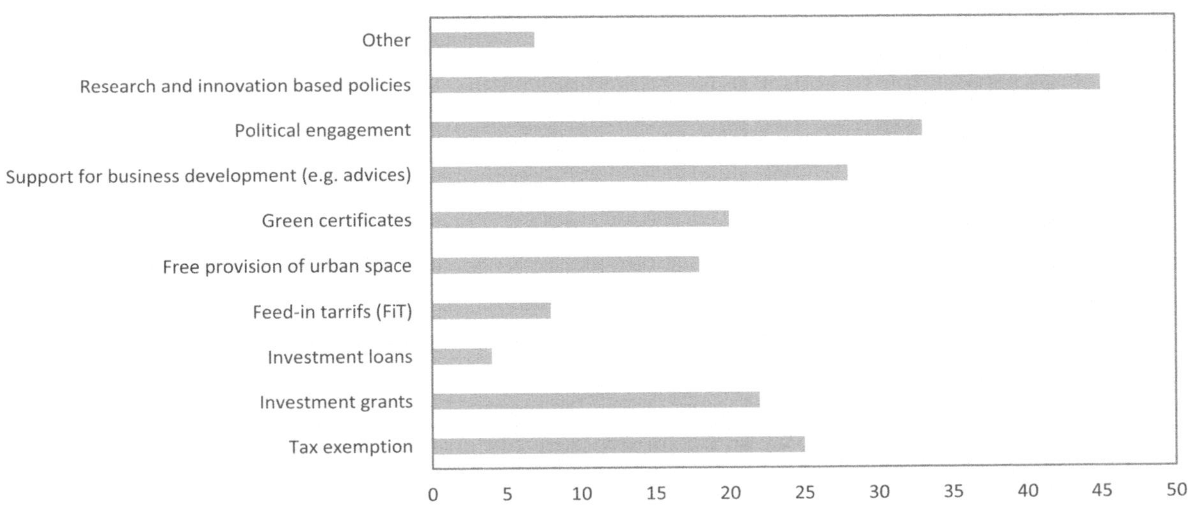

Figure 5 | Supporting measures considered by projects.

Stakeholder engagement and awareness

To understand the degree of stakeholder awareness and engagement in practice, the following analysis was conducted. Findings related to the types of stakeholders involved in the different projects are presented in Figure 6. In total 12 types of involved stakeholders were identified in the projects, with

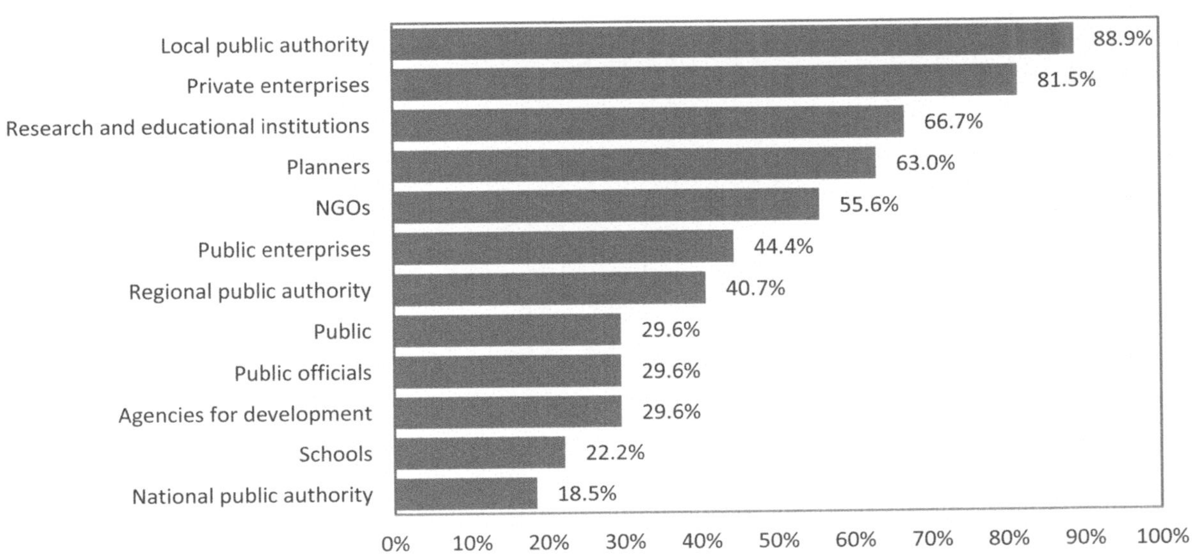

Figure 6 | Types of stakeholders involved in the projects.

public authority bodies and private enterprises having the lion's share of participation while individual citizens are the least represented, confirming the lack of horizontal acceptance of NBS. NBS as a concept is multidimensional and its implementation requires collaboration amongst different policy areas, sectors and stakeholders (Van Ham & Klimmek 2017). More precisely, Figure 6 shows that the most represented type of stakeholder identified in the analysed projects was Local public authority (88.9%). Private enterprises hold the second place (81.5%), followed by Research and educational institutions (66.7%) and Planners (63%) while, on the place are National public authorities (18.5%).

Regarding the barriers for the implementation of project activities related to stakeholder awareness and engagement in the analysed projects (respondents could choose multiple answers), unsupportive legal frameworks (45.8%) and insufficient financial resources (45.8%) are identified as the main issues (Figure 7). These findings are further supported by literature, e.g. Brink *et al.* (2016) indicate that the lack of resources, know-how, tools, unsupportive legal frameworks are, among others, serious impediments to NBS implementation. Lack of time holds third place with 33.3%, followed by lack of basic knowledge about NBS (29.2%) and Lack of engagement plan (25%). Combination of the previously mentioned five barriers for stakeholder awareness and engagement is present in almost every project.

Figure 8 presents the most common participation tools and techniques that are used in the projects for stakeholder awareness raising and engagement on NBS. Among others, respondents identified Participatory workshops (63.6%), Internet (45.5%), Public meetings (45.5%), Printed information (e.g. brochures, leaflets, newsletters) (40.9%) and Regional focus groups events (27.3%) are the most common participation tools and techniques used for stakeholder awareness and engagement on NBS. Besides these five categories, there were also external events (e.g. fairs, promotions, exhibitions), telephone contact, newspaper and semi-structured interviews, represented with less than 25% in provided answers. Results from Figure 10 indicate a shift from 'traditional techniques and tools', such as telephone and printed media, to the increased use of social media and the internet. The fact that social media and internet are more popular probably also depends on the type of stakeholders that are commonly engaged in participatory events (youth and urbanized population), while an older population favours printed media. It is important to consider that a 'participation divide' (Hargittai & Walejko 2008) between elders and youth, urban and rural, and middle-high and low income affects such a result (Hargittai 2002; Paul & Stegbauer 2005; Sylvester & McGlynn 2010). Further research might be useful for future NBS implementation considering different groups of stakeholders.

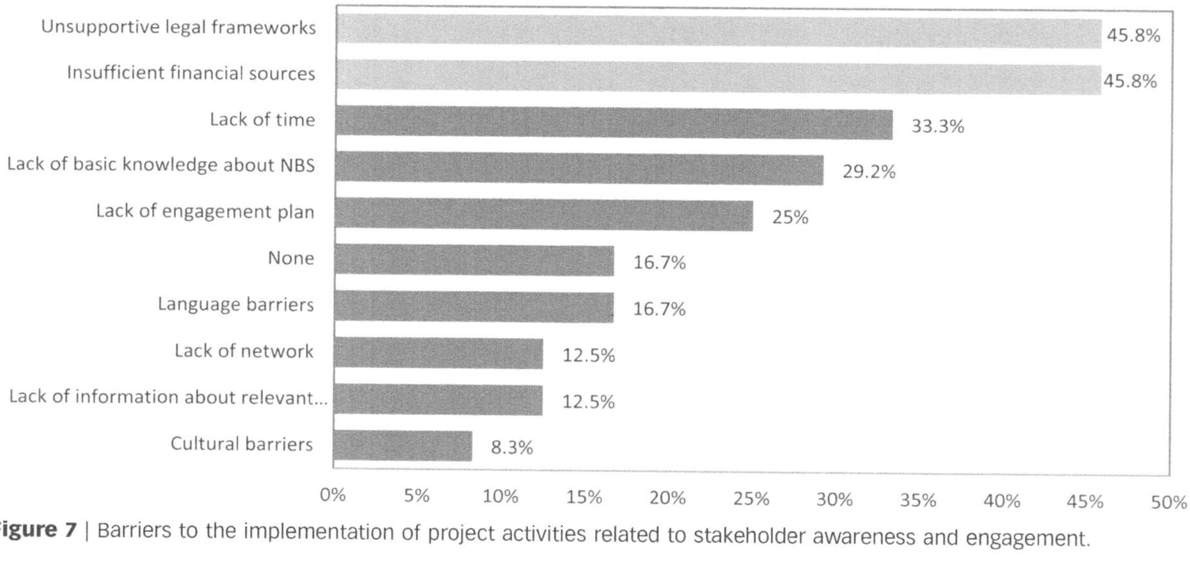

Figure 7 | Barriers to the implementation of project activities related to stakeholder awareness and engagement.

doi: 10.2166/bgs.2020.929

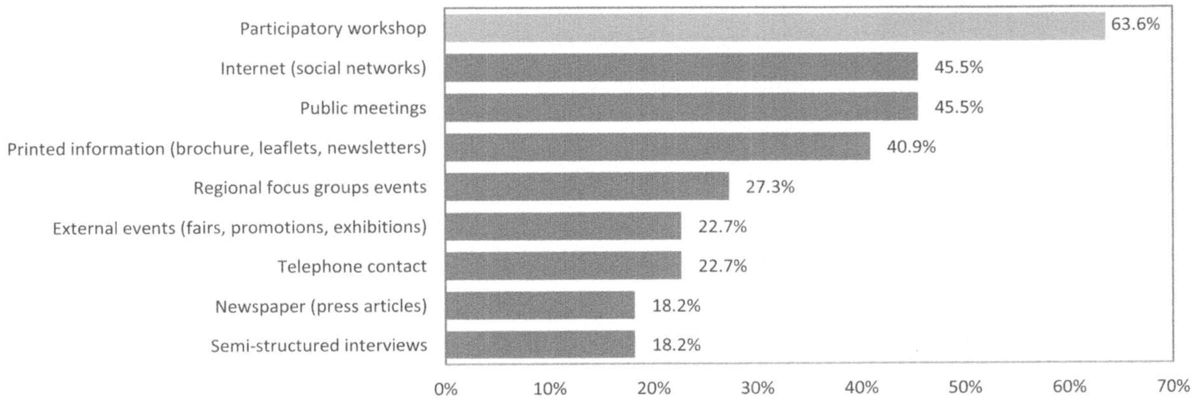

Figure 8 | Public participation tools and techniques used in the project for stakeholder awareness raising and engagement on NBS.

According to the International Association for Public Participation, there are five levels of stakeholder engagement, i.e. inform, consult, involve, collaborate and empower (International Association for Public Participation (IAP2) 2007). According to this categorization, the level of stakeholder engagement in the analysed projects was identified (Figure 9). It should be noted though that the results only represent the respondents' perception regarding the level of stakeholder engagement in their project.

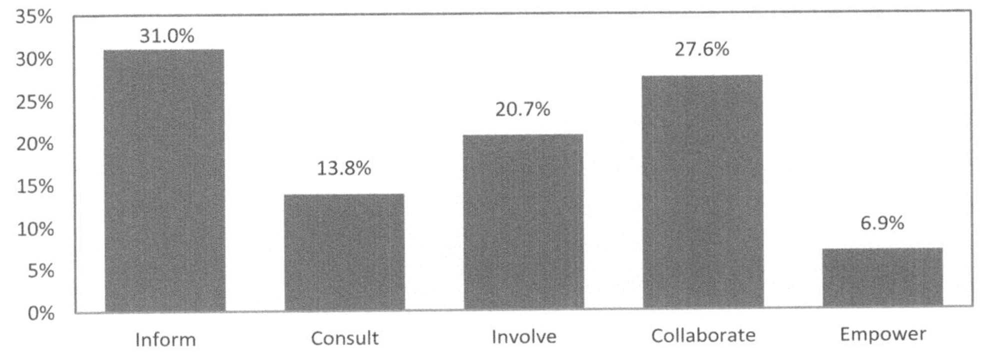

Figure 9 | Level of stakeholder engagement in the projects.

In the majority of the projects (31%), stakeholders were informed in order to understand the problem, alternatives, and opportunities related to NBS. In 13.8% of the projects, stakeholders had a consulting role in order to provide feedback on the analysis, the alternatives and/or the decisions related to NBS. 20.7% of the respondents involved stakeholders in their projects to work directly with them throughout the process in order to ensure that stakeholders' concerns and aspirations about NBS are consistently understood and considered. In 27.6% of the projects, collaborative methods (e.g. collaboration for the development of alternatives, identification of preferred solutions) – which help mobilize stakeholders and build capacity to deliver projects (Healey 1998) – are deployed in stakeholder engagement. Regarding the highest level of stakeholder engagement, only 6.9% of the respondents empowered stakeholders in their project to make a final decision.

Results of cross-tabulation analysis between the different types of the projects and level of stakeholder engagement are presented in Figure 10. The results of this analysis reveal that Research and Innovation projects engage the highest number of stakeholders and among them, in 20.7% of the R&I projects stakeholders were informed, in 10.3% they were consulted, in 10.3% they were involved, in 13.8% they collaborated and in 6.9% stakeholders were empowered.

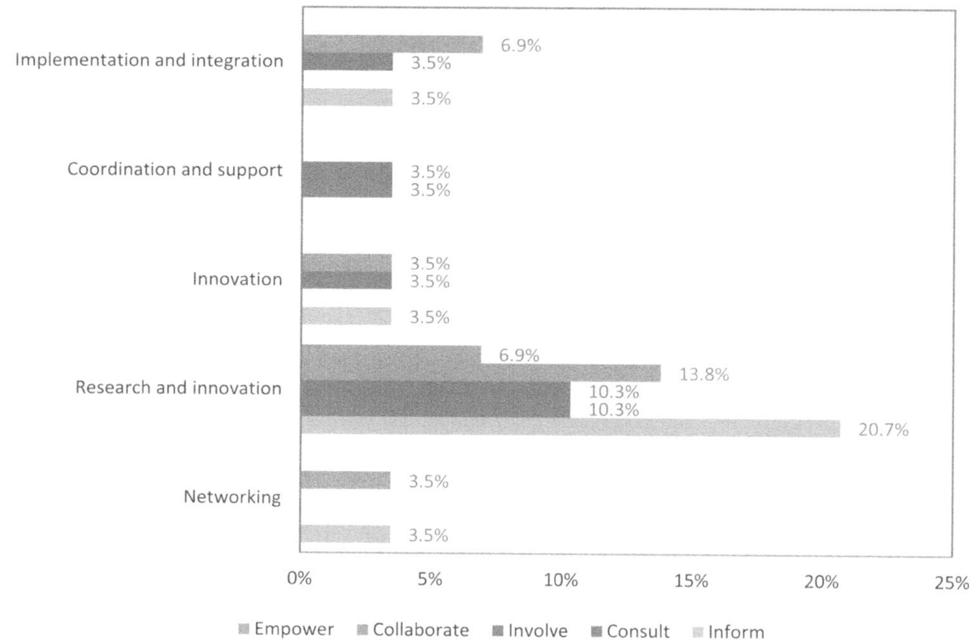

Figure 10 | Cross-tabulation between the type of the project and level of engagement.

An additional cross-tabulation analysis between the level of project's implementation and the level of stakeholder engagement (Figure 11) reveals that projects implemented at the macro level (city, regional) engaged the highest number of stakeholders out of which, 20.69% were informed, 6.9% were consulted, 10.34% of the stakeholders were involved, 10.34% were collaborated and 3.45% were empowered.

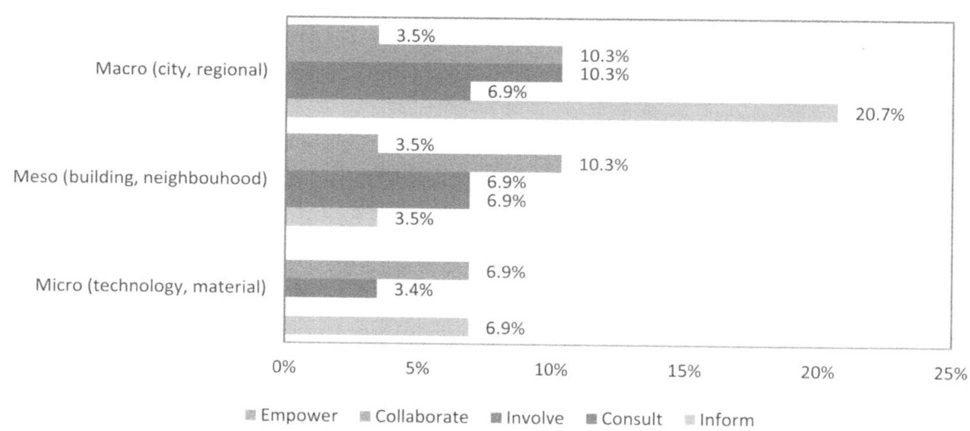

Figure 11 | Cross-tabulation between application scale of the project and level of engagement.

Tools and methods

To understand which assessment methodologies are mostly employed in practice, the following analysis was conducted. The assessment methodologies that are applied in the targeted projects are presented in Figure 12, indicating that a wide variety of assessment methodologies is used. In general, it is worth noting that the assessment of NBS is mostly focused on the environmental and economic aspects, while the social aspect of the NBS implementation is underestimated (i.e. 12.5% of the

doi: 10.2166/bgs.2020.929

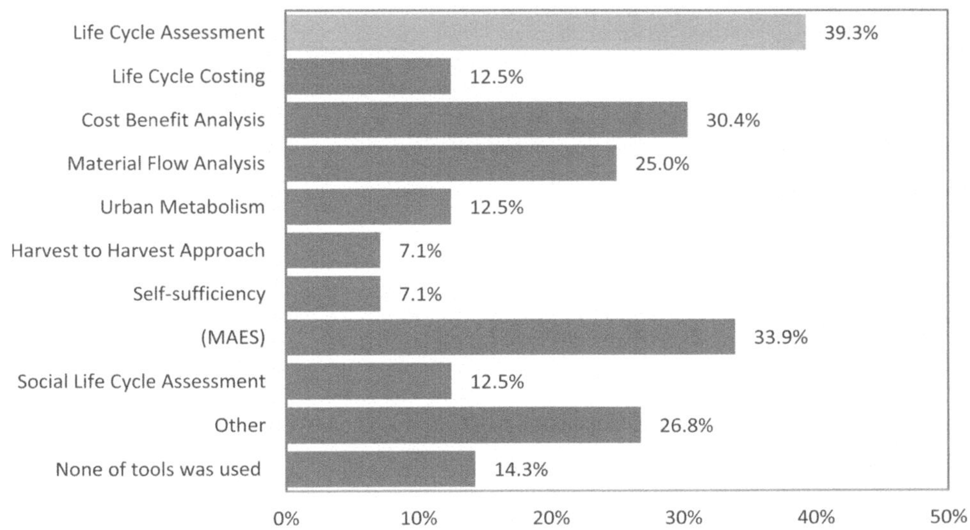

Figure 12 | Methods used for the assessment of the applied solutions.

projects conducted S-LCA) in the projects. More precisely, most projects apply one of the following assessment methodologies: (i) LCA (39.3%), a well-established method for sustainability assessment; (ii) Cost Benefit Analysis (30.4%), in order to prove the cost effectiveness of the implemented solutions; and (iii) Mapping and Assessment of the Ecosystem Services (MAES) (33.9%), as the enhancement of the ecosystem services is the cornerstone of NBS. MFA is also considerably applied in the projects (25%), which can be explained by the fact that it is another well-established method to measure the circularity of systems (EC 2015b; Linder *et al.* 2017). Consequently, it can be argued that in practice (i.e. implementation of a project) well-established methods are more favourable compared to other newly developed methods (e.g. HHA or SS) or to methods that are not directly linked to NBS or circularity assessment (i.e. Urban Metabolism).

Regarding the application scale of the projects (Figure 3(c)), several projects were carried out on-ground. As those projects might be very useful for the transferability of the concepts of NBS for Circular Cities, a further analysis of the results of on-ground projects regarding their application scale was carried out. Figure 13 shows that 52.6% of these on-ground projects have been applied on a macro-scale, and 39.5% of them have been applied on a meso-scale.

The methodologies that have been used in on-ground projects to assess the effectiveness of the applied technologies with regard to their application scale are summarized in Figure 14 – (left) assessment methodologies applied on a meso-scale, and (right) assessment methodologies that have been applied on a macro-scale.

Meso-scale projects (Figure 14-left) mostly use Life Cycle Assessment (LCA) (46.7%) to assess the sustainability of their proposed technologies, while 40% of these projects additionally use material flow analysis (MFA). It is evident that the economic aspect is underestimated among these projects, since 20% use Cost Benefit Analysis (CBA) and only 6.7% use Life Cycle Costing (LCC). In contrast,

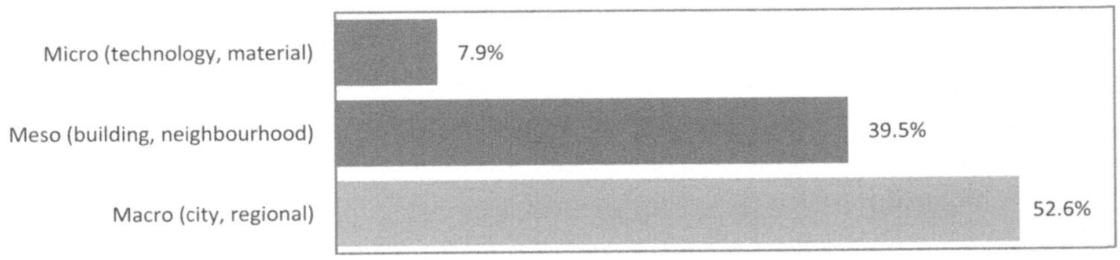

Figure 13 | Distribution of on-ground projects regarding their application scale.

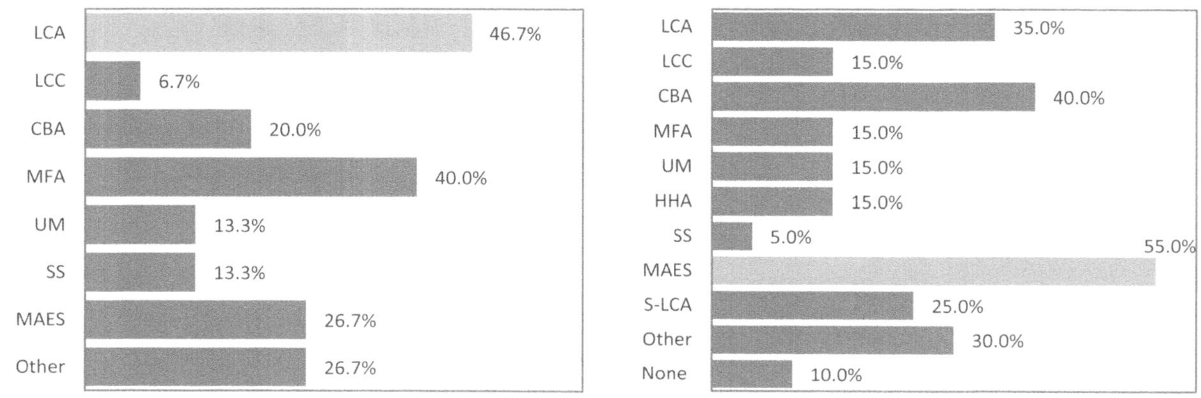

Figure 14 | Assessment methods for on-ground projects on meso- (left) and macro-scale (right).

this trend changes significantly when on-ground projects applied on a macro-scale. In this case (Figure 14-right), the most widely applied methodology is the Mapping and Assessment of Ecosystem Services (MAES) and only 35% of the projects use LCA. Ecosystem Services are implemented in larger scales (e.g. catchment-scale, watershed-scale, etc.), while in a smaller scale (e.g. neighbourhood scale) only the relative changes in ecosystem services can be assessed. LCA on the other hand, is mostly used to assess the sustainability of specific systems (meso-scale), e.g. wastewater treatment plants. Although LCA can be applied at larger scales as well, there are only a few studies that have performed this type of analysis. For example the study of Lane *et al.* (2015) includes the assessment of the environmental impact of a broad range of technologies at the 'whole-of-system' level – urban water system. Therefore, the change in the preference of the assessment methodologies would be better explained if the assessment scale is also known. In case the assessment scale coincides with the application scale of the technologies, then the reason behind this change in preference is clear. Moreover, on-ground projects at macro-scale are more focused on the economic aspect of their technologies, as 40% of these projects use CBA.

Some useful ICT tools were used and reported in these projects too. They can be grouped as (i) data or databases (including GIS for 3D visualization, weather gridded, energy carbon footprint, data mining sensors, cloud-based geo-referenced system for storing and communicating the acquired water quality information, or websites containing data for mapping of pilot areas), (ii) models (for temporary housing, climate adaptation, water management strategies or for gaming in some apps), (iii) monitoring and control systems (for online greenhouse gases emission in wastewater treatment plants based on wireless sensor networks, for ecosystem services adapted to urban areas, for real-time flood mitigation, for waterloops or for automation and process control in general), (iv) Decision Support Systems to select among alternatives or to identify potential pollutant sources and predict the effectiveness of mitigation measures. Commercial software (e.g. simulistics, UnicaLids or MATLAB, etc.) and platforms with open-source information (e.g. OPPLA, thinknature, Tygron, ICT governance, or for e-learning courses like edXMOOCs) have also been reported for the implementation of NBS.

CONCLUSIONS

Adoption and implementation of NBS in Circular Cities is circular itself and require four main steps, i.e., planning, design, assessment, and communication. Policies and regulations, stakeholder engagement and awareness, and tools and methods assessing the socio-economic and environmental impacts of the solutions are integral parts of this circular process.

The state-of-the-art analysis performed in the present paper, revealed that limited research has been conducted on 'policies, regulations and governance' for deploying NBS to move towards Circular

Cities; a deeper analysis is still required. Circularity initiatives and NBS imply risks and unknowns, compared to classical solutions, and are not compatible with current rules and regulations. The survey results confirmed these findings revealing that many of the current policies and regulations are almost equally perceived as both limiting and enabling the development of NBS. Experimentation zones where these prevention principles and regulations are not (fully) applicable for specific projects could be a solution. The EU is moving in that direction with the 'innovation deals' initiative, in particular one launched on treated wastewater reuse.

The literature revealed that public awareness and social acceptance are key issues related to the success of NBS for Circular Cities, as they can reduce barriers to NBS adoption and diffusion on a wider scale of application. Respectively, the projects' survey demonstrated that the level of stakeholder engagement is very important in order to achieve accepted outcomes, as well as for successful project delivery. However, unsupportive legal frameworks and insufficient financial sources were identified as the main barriers related to stakeholder awareness and engagement. The projects' survey demonstrated that participatory workshops, social media and public meetings are the most common public participation tools and techniques used in the projects for raising stakeholder awareness and engagement. Therefore, this study implies that such tools should be further investigated with regards to each one of the four steps for NBS application so that their utilization is more productive.

The analysis of existing tools advocates more quantitative upscaling of NBS technologies, although research and innovation in this field seems to be still in their infancy. An extensive diversity of tools and methods assessing the impacts of NBS was identified in the literature, as well as in the selected projects, which complicates the comparability and measurability of such projects, as well as their transferability on a wider scale. However, based on the results from the survey it was found that the assessment methods that are mostly used are the ones that are well-established, such as LCA, CBA, MFA, while recently developed methodologies (e.g. HHA) are not in favour even though they may be very prominent for the assessment of NBS for creating Circular Cities. Interestingly, it was found that the decision on the employed methodologies is related to the application scale of the project. A further and more focused analysis on the nature of tools applied for every step of the application of NBS, i.e., planning (routinely considering NBS, integrate triple benefits targets: economy, community and environment, build new partnerships to bring new resources and skills), design (using advanced tools and guidelines integrating NBS), assessment (learning new lessons for closing the knowledge gap, and communication (engaging policy makers, building capacities), would lead to a more straightforward methodology. More specifically, the development of a widely accepted methodology or framework for assessing NBS for Circular Cities would provide the guidelines regarding the hybridization of the different methods and it would systematize the evaluation of the effectiveness of NBS for creating Circular Cities.

Improving knowledge on the impact of NBS is moreover, necessary for decision makers to prepare a transition process to circular systems. The EU is leading a wide range of developments in particular with enabling environments (e.g. policies, innovation funds), while at the global level, IUCN (IUCN 2012; Cohen-Schacham *et al.* 2016) and the World Bank (WB 2019) are advocating for NBS to be more integrated into new initiatives, in particular infrastructure projects.

Continuous monitoring that would be enabled through the use of ICT tools and evaluation of the implementation of NBS for circular cities will support the development of a solid knowledge base for more suitable policies and regulations.

ACKNOWLEDGEMENTS

The work is funded within the COST Action CA17133 Circular City ('Implementing nature based solutions for creating a resourceful circular city', http://www.circular-city.eu, 22 October 2018–21

October 2022). COST Actions are funded within the EU Horizon 2020 Programme. The authors are grateful for the support. Our gratitude to Cecília Delgado, from NOVA School of Social Sciences and Humanities and CICS.NOVA – *Universidade NOVA de Lisboa, Portugal,* for her contribution to this paper on Food Policy Councils stakeholder engagement tools.

DISCLAIMER

The content of this article is the authors' responsibility and neither COST nor any person acting on its behalf is responsible for the use, which might be made of the information contained in it.

SUPPLEMENTARY MATERIAL

The Supplementary Material for this paper is available at https://dx.doi.org/10.2166/bgs.2020.929.

REFERENCES

Alberti, M. 2008 *Advances in Urban Ecology: Integrating Humans and Ecological Processes in Urban Ecosystems.* Springer Science + Business Media, LLC, New York, NY, USA.

Andrews, E. S., Barthel, L.-P., Beck, T., Benoit, C., Ciroth, A., Cucuzzella, C., Gensch, C.-O., Hubert, J., Lesage, P., Manhart, A., Maria, L., Mazeau, P., Mazijn, B., Methot, A.-L., Moberg, A., Norris, G. A., Parent, J., Prakash, S., Reveret, J.-P., Spillemaeckers, S., Ugaya, S., Valdivia, S. & Weidema, B. P. 2009 *Guidelines for Social Life Cycle Assessment of Products.* UNEP/SETAC Life Cycle Initiative.

Bach, P. M., McCarthy, D. T. & Deletic, A. 2015 Can we model the implementation of water sensitive urban design in evolving cities? *Water Science and Technology* **71** (1), 149–156.

Bassarab, K., Santo, R. & Palmer, A. 2019 *Food Policy Council Report 2018.* Johns Hopkins University Center for a Livable Future, Baltimore, MD, USA.

Bastein, A. G. T. M., Verstraeten-Jochemsen, J. N., Rietveld, E., Hauck, M., Frijters, E., Klijn, O. & Driessen, B. 2016 *Circular Amsterdam. A Vision and Action Agenda for the City and Metropolitan Area.* TNO.

Beaussier, T., Caurla, S., Bellon-Maurel, V. & Loiseau, E. 2019 Coupling economic models and environmental assessment methods to support regional policies: a critical review. *Journal of Cleaner Production* **216**, 408–421.

Beck, F. & Martinot, E. 2004 *Renewable Energy Policies and Barriers.*

Beloin-Saint-Pierre, D., Rugani, B., Lasvaux, S., Mailhac, A., Popovici, E., Sibiude, G., Benetto, E. & Schiopu, N. 2017 A review of urban metabolism studies to identify key methodological choices for future harmonization and implementation. *Journal of Cleaner Production* **163**, S223–S240.

Binder, T., Ehn, P., De Michelis, M., Jacucci, G., Linde, P. & Wagner, I. 2011 *Design Things.* MIT Press, Cambridge, MA.

Brink, E., Aalders, T., Ádám, D., Feller, R., Henselek, Y., Hoffmann, A., Ibe, K., Matthey-Doret, A., Meyer, M., Negrut, N. L., Rau, A.-L., Riewerts, B., von Schuckmann, L., Törnros, S., von Wehrden, H., Abson, D. J. & Wamsler, C. 2016 Cascades of green: a review of ecosystem-based adaptation in urban areas. *Global Environmental Change* **36**, 111–123.

Brunner, P. H. & Rechberger, H. 2016 *Handbook of Material Flow Analysis: for Environmental, Resource, and Waste Engineers.* CRC Press.

Calliari, E., Staccione, A. & Mysiak, J. 2019 An assessment framework for climate-proof nature-based solutions. *Science of The Total Environment* **656**, 691–700.

Carrera, D. G. & Mack, A. 2010 Sustainability assessment of energy technologies via social indicators: results of a survey among European energy experts. *Energy Policy* **38** (2), 1030–1039.

Chester, M., Pincetl, S. & Allenby, B. 2012 Avoiding unintended tradeoffs by integrating life-cycle impact assessment with urban metabolism. *Current Opinion in Environmental Sustainability* **4** (4), 451–457.

Cohen-Schacham, E., Walters, G., Janzen, C., Maginnis, S. 2016 Nature-based solutions: from theory to practice. (Iucn, ed.), *Nature-based Solutions to Address Global Societal Challenges.* IUCN, Galnd, Switzerland, pp. xiii + 97.

COM 2005 *Thematic Strategy on the Urban Environment.*

COM 2011 *EU Biodiversity Strategy to 2020.*

COM 2013 *Green Infrastructure.* 249 final.

COM 2019 *Report on the Implementation of the Circular Economy Action Plan.* SWD(2019) 90 final.

Connor, R. 2015 *The United Nations World Water Development Report 2015: Water for A Sustainable World.* UNESCO Publishing.

EC 2015a *Towards an EU Research and Innovation Policy Agenda for Nature-Based Solutions & Re-Naturing Cities*. EC, Brussels, Belgium.

EC 2015b *Communication From the Commission to the European Parliament, the Council, the European Economic and Social Committee and the Committee of the Regions. Closing the Loop – An EU Action Plan for the Circular Economy*. European Commission, Brussels.

Frantzeskaki, N., Wittmayer, J. & Loorbach, D. 2014 The role of partnerships in 'realising' urban sustainability in Rotterdam's City Ports Area, The Netherlands. *Journal of Cleaner Production* **65**, 406–417.

Glasgow Chamber of Commerce 2016 *Zero Waste Scotland; Glasgow City Council; Circle Economy. Circular Glasgow: A Vision and Action Plan for the City of Glasgow*. Glasgow Chamber of Commerce, Glasgow. Available from: https://www.circle-economy.com/wp-content/uploads/2016/06/circular-glasgow-report-weblow-res.pdf.

Goldstein, B., Birkved, M., Quitzau, M. B. & Hauschild, M. 2013 Quantification of urban metabolism through coupling with the life cycle assessment framework: concept development and case study. *Environmental Research Letters* **8**, 035024.

González-Val, R. & Pueyo, F. 2019 Natural resources, economic growth and geography. *Economic Modelling*., Elsevier, vol. 83(C), pp. 150–159.

Hargittai, E. 2002 Second-level digital divide: differences in people's online skills. *First Monday* **7** (4). https://doi.org/10.5210/fm.v7i4.942.

Hargittai, E. & Walejko, G. 2008 The participation divide: content creation and sharing in the digital age. *Information, Community and Society* **11** (2), 239–256.

Healey, P. 1998 Building institutional capacity through collaborative approaches to urban planning. *Environment and Planning A* **30** (9), 1531–1546.

International Association for Public Participation (IAP2) 2007 *IAP2 Public Participation Spectrum*. Available from: https://www.iap2.org.au/Tenant/C0000004/00000001/files/IAP2_Public_Participation_Spectrum.pdf.

ISO 14040 2006 *Environmental Management-Life Cycle Assessment-Principles and Framework*. International Organization for Standardization, Geneva, Switzerland.

IUCN 2012 *The IUCN Programme 2013–2016*. IUCN, Gland, p. 30.

Kabisch, N., Frantzeskaki, N., Pauleit, S., Naumann, S., Davis, M., Artmann, M., Haase, D., Knapp, S., Korn, H., Stadler, J. & Zaunberger, K. 2016 Nature-based solutions to climate change mitigation and adaptation in urban areas: perspectives on indicators, knowledge gaps, barriers, and opportunities for action. *Ecology and Society* **21** (2), 39.

Kennedy, C., Cuddihy, J. & Engel-Yan, J. 2007 The changing metabolism of cities. *Journal of Industrial Ecology* **11**, 43–59.

Lane, J. L., de Haas, D. W. & Lant, P. A. 2015 The diverse environmental burden of city-scale urban water systems. *Water Research* **81**, 398–415.

Langergraber, G., Pucher, B., Simperler, L., Kisser, J., Katsou, E., Buehler, D., Garcia Mateo, M. C. & Atanasova, N. 2020 Implementing nature-based solutions for creating a resourceful circular city. *Blue-Green Systems* **2** (1), 173–185.

Linder, M., Sarasini, S. & van Loon, P. 2017 A metric for quantifying product-level circularity. *Journal of Industrial Ecology* **21** (3), 545–558.

Liquete, C., Udias, A., Conte, G., Grizzetti, B. & Masi, F. 2016 Integrated valuation of a nature-based solution for water pollution control. Highlighting hidden benefits. *Ecosystem Services* **22**, 392–401.

Maes, J. & Jacobs, S. 2017 Nature-based solutions for Europe's sustainable development. *Conservation Letters* **10**, 121–124. http://dx.doi.org/10.1111/conl.12216.

Maes, J., Teller, A., Erhard, M., Liquete, C., Braat, L., Berry, P., Egoh, B., Puydarrieux, P., Fiorina, C., Santos, F., Paracchini, M. L., Keune, H., Wittmer, H., Hauck, J., Fiala, I., Verburg, P. H., Condé, S., Schägner, J. P., San Miguel, J., Estreguil, C., Ostermann, O., Barredo, J. I., Pereira, H. M., Stott, A., Laporte, V., Meiner, A., Olah, B., Royo Gelabert, E., Spyropoulou, R., Petersen, J. E., Maguire, C., Zal, N., Achilleos, E., Rubin, A., Ledoux, L., Brown, C., Raes, C., Jacobs, S., Vandewalle, M., Connor, D. & Bidoglio, G. 2013 *Mapping and Assessment of Ecosystems and Their Services. An Analytical Framework for Ecosystem Assessments Under Action 5 of the EU Biodiversity Strategy to 2020*. Publications office of the European Union, European Union, Luxembourg.

Maia de Souza, D., Russo Lopes, G., Hansson, J. & Hansen, K. 2018 Ecosystem services in life cycle assessment: a synthesis of knowledge and recommendations for biofuels. *Ecosystem Services* **30**, 200–210.

Mairie de Paris. 2017 *Paris Circular Economy Plan*.

Marin, J. & De Meulder, B. 2018 Interpreting circularity. Circular city representations concealing transition drivers. *Sustainability* **10**, 1310.

Marvuglia, A., Rugani, B. & Adell, G. 2018 Designing sustainable technologies, products and policies: from science to innovation. In: *8th International Conference on Life Cycle Management (LCM), LCM at the Urban Scale: BIM and Nature Based Solutions*. Springer International Publishing, Luxembourg.

Mazé, R. 2007 *Occupying Time: Design, Time, and the Form of Interaction. PhD*, Blekinge Institute of Technology, Blekinge.

McIntosh, B. S., Ascough, J. C., Twery, M., Chew, J., Elmahdi, A., Haase, D., Harou, J., Hepting, D., Cuddy, S., Jakeman, A. J., Chen, S., Kassahun, A., Lautenbach, S., Matthews, K., Merritt, W., Quinn, N. W. T., Rodriguez-Roda, I., Sieber, S., Stavenga, M., Sulis, A., Ticehurst, J., Volk, M., Wrobel, M., van Delden, H., El-Sawah, S., Rizzoli, A. & Voinov, A. 2011 Environmental decision support systems (EDSS) development – challenges and best practices. *Environmental Modelling & Software* **26** (12), 1389–1402. doi:10.1016/j.envsoft.2011.09.009.

Morales-Torres, A., Escuder-Bueno, I., Andrés-Doménech, I. & Perales-Momparler, S. 2016 Decision support tool for energy-efficient, sustainable and integrated urban stormwater management. *Environmental Modelling & Software* **84**, 518–528.

Naturvation 2017 *The Governance and Politics of Nature-Based Solutions.*

Nesshöver, C., Assmuth, T., Irvine, K. N., Rusch, G. M., Waylen, K. A., Delbaere, B., Haase, D., Jones-Walters, L., Keune, H., Kovacs, E., Krauze, K., Külvik, M., Rey, F., van Dijk, J., Vistad, O. I., Wilkinson, M. E. & Wittmer, H. 2017 The science, policy and practice of nature-based solutions an interdisciplinary perspective. *Science of The Total Environment* **579** (1), 1215–1227.

Niță, M.-R., Anghel, A.-M., Bănescu, C., Munteanu, A.-M., Pesamosca, S.-S., Zețu, M. & Popa, A.-M. 2017 Are Romanian urban strategies planning for green? *European Planning Studies* 1–16. http://dx.doi.org/10.1080/09654313.2017.1382446. 10.1080/09654313.2017.1382446.

Onat, N., Kucukvar, M., Halog, A. & Cloutier, S. 2017 Systems thinking for life cycle sustainability assessment: a review of recent developments, applications, and future perspectives. *Sustainability* **9**, 706.

Othoniel, B., Rugani, B., Heijungs, R., Benetto, E. & Withagen, C. 2016 Assessment of life cycle impacts on ecosystem services: promise, problems, and prospects. *Environmental Science & Technology* **50**, 1077–1092.

Paul, G. & Stegbauer, C. 2005 Is the digital divide between young and elderly people increasing? *First Monday* **10** (10).

Pearce, D. 1998 Cost-benefit analysis and environmental policy. *Oxford Review of Economic Policy* **14** (4), 84–100.

Petit-Boix, A. & Leipold, S. 2018 Circular economy in cities: reviewing how environmental research aligns with local practices. *Journal of Cleaner Production* **195**, 1270–1281.

Petit-Boix, A., Llorach-Massana, P., Sanjuan-Delmás, D., Sierra-Pérez, J., Vinyes, E., Gabarrell, X., Rieradevall, J. & Sanyé-Mengual, E. 2017 Application of life cycle thinking towards sustainable cities: a review. *Journal of Cleaner Production* **166**, 939–951.

Pickett, S. T., Cadenasso, M. L., Grove, J. M., Groffman, P. M., Band, L. E., Boone, C. G., Burch, W. R., Grimmond, C. S. B., Hom, J. & Jenkins, J. C. 2008 Beyond urban legends: an emerging framework of urban ecology, as illustrated by the Baltimore Ecosystem Study. *BioScience* **58**, 139–150.

Pincetl, S., Bunje, P. & Holmes, T. 2012 An expanded urban metabolism method: toward a systems approach for assessing urban energy processes and causes. *Landscape and Urban Planning* **107**, 193–202.

Poch, M., Comas, J., Cortés, U., Sànchez-Marrè, U. & Rodríguez-Roda, I. 2017 Crossing the death valley to transfer environmental decision support systems to the water market. *Global Challenges* **1** (3). doi:10.1002/gch2.201700009.

Potschin-Young, M., Haines-Young, R., Görg, C., Heink, U., Jax, K. & Schleyer, C. 2018 Understanding the role of conceptual frameworks: reading the ecosystem service cascade. *Ecosystem Services* **29**, 428–440. Available from: http://www.sciencedirect.com/science/article/pii/S221204161630523X. https://doi.org/10.1016/j.ecoser.2017.05.015.

Prendeville, S., Cherim, E. & Bocken, N. 2018 Mapping six cities in transition. *Environmental Innovation and Societal Transitions* **26**, 171–194.

Radinja, M., Comas, J., Corominas, L. & Atanasova, N. 2019 Assessing stormwater control measures using modelling and a multi-criteria approach. *Journal of Environmental Management* **243**, 257–268. https://doi.org/10.1016/J.JENVMAN.2019.04.102.

Raymond, C. M., Berry, P., Breil, M., Nita, M. R., Kabisch, N., de Bel, M., Enzi, V., Frantzeskaki, N., Geneletti, D., Cardinaletti, M., Lovinger, L., Basnou, C., Monteiro, A., Robrecht, H., Sgrigna, G., Munari, L. & Calfapietra, C. 2017a *An Impact Evaluation Framework to Support Planning and Evaluation of Nature-Based Solutions Projects.* Centre for Ecology & Hydrology, EKLIPSE Expert Working Group, Wallingford, United Kingdom.

Raymond, C. M., Frantzeskaki, N., Kabisch, N., Berry, P., Breil, M., Nita, M. R., Geneletti, D. & Calfapietra, C. 2017b A framework for assessing and implementing the co-benefits of nature-based solutions in urban areas. *Environmental Science & Policy* **77**, 15–24. Available from: http://www.sciencedirect.com/science/article/pii/S1462901117306317. http://dx.doi.org/10.1016/j.envsci.2017.07.008

Redström, J. 2008 RE: Definitions of use. *Design Studies* **29** (4), 410–423.

Rincon, L., Castell, A., Perez, G., Sole, C., Boer, D. & Cabeza, L. F. 2013 Evaluation of the environmental impact of experimental buildings with different constructive systems using. *Material Flow Analysis and Life Cycle Assessment* **109**, 544–552.

Rizzo, F., Deserti, A. & Cobanli, O. 2015 Design and social innovation for the development of human smart cities. *Nordes* **1** (6).

Roddis, P., Carver, S., Dallimer, M., Norman, P. & Ziv, G. 2018 The role of community acceptance in planning outcomes for onshore wind and solar farms: an energy justice analysis. *Applied Energy* **226**, 353–364.

Rödger, J. M., Kjær, L. L. & Pagoropoulos, A. 2018 Life cycle costing: an introduction. In: *Life Cycle Assessment* (J. M. Rödger, L. L. Kjær & A. Pagoropoulos, eds). Springer, Cham, pp. 373–399.

Rosso-Cerón, A. M. & Kafarov, V. 2015 Barriers to social acceptance of renewable energy systems in Colombia. *Current Opinion in Chemical Engineering* **10**, 103–110.

Sack-Nielsen, T. 2018 *Circularity City Book: Shaping our Urban Future.* KLS PurePrint, Denmark.

Schaubroeck, T. 2018 Towards a general sustainability assessment of human/industrial and nature-based solutions. *Sustainability Science* **13** (4), 1185–1191.

Schön, J. 1983 *Petrophysik: PhysikalischeEigenschaften von Gesteinen und Mineralen.* Akademie-Verlag, Berlin.

Sohn, J., Vega, G. C. & Birkved, M. 2018 A methodology concept for territorial metabolism – life cycle assessment: challenges and opportunities in scaling from urban to territorial assessment. *Procedia CIRP* **69**, 89–93.

Stewart, R., Bey, N. & Boks, C. 2016 Exploration of the barriers to implementing different types of sustainability approaches. *Procedia CIRP* **48**, 22–27.

Sylvester, D. E. & McGlynn, A. J. 2010 The digital divide, political participation, and place. *Social Science Computer Review* **28** (1), 64–74.

Ugolini, F., Massetti, L., Sanesi, G. & Pearlmutter, D. 2015 Knowledge transfer between stakeholders in the field of urban forestry and green infrastructure: results of a European survey. *Land Use Policy* **49**, 365–381.

UN 2018 *UN World Water Development Report, Nature-Based Solutions for Water.* UN.

Van Ham, C. & Klimmek, H. 2017 Partnerships for nature-based solutions in urban areas. In: *Nature-Based Solutions to Climate Change Adaptation in Urban Areas. Theory and Practice of Urban Sustainability Transitions* (Kabisch, N., Korn, H., Stadler, J. & Bonn, A., eds). Springer, Germany.

Verones, F., Bare, J., Bulle, C., Frischknecht, R., Hauschild, M., Hellweg, S., Henderson, A., Jolliet, O., Laurent, A., Liao, X., Lindner, J. P., Maia de Souza, D.,, Michelsen, O., Patouillard, L., Pfister, S., Posthuma, L., Prado, V., Ridoutt, B., Rosenbaum, R. K., Sala, S., Ugaya, C., Vieira, M. & Fantke, P. 2017 LCIA framework and cross-cutting issues guidance within the UNEP-SETAC Life Cycle Initiative. *Journal of Cleaner Production* **161**, 957–967.

Voskamp, I. M. & Van de Ven, F. H. M. 2015 Planning support system for climate adaptation: composing effective sets of blue-green measures to reduce urban vulnerability to extreme weather events. *Building and Environment* **83**, 159–167. http://dx.doi.org/10.1016/j.buildenv.2014.07.018.

Walker, G. 2009 Beyond distribution and proximity: exploring the multiple spatialities of environmental justice. *Antipode* **41** (4), 614–636.

WB 2019 *World Bank 2019, Integrating Green and Gray: Creating Next Generation Infrastructure.*

WCED Brundtland Commission 1987 *Our Common Future.* Oxford University Press, New York.

Wielemaker, R. C., Weijma, J. & Zeeman, G. 2018 Harvest to harvest: recovering nutrients with new sanitation systems for reuse in urban agriculture. *Resources, Conservation and Recycling* **128**, 426–437.

Williams, J. 2019 Circular cities: challenges to implementing looping actions. *Sustainability* **11**, 423.

Wüstenhagen, R., Wolsink, M. & Bürer, M. J. 2007 Social acceptance of renewable energy innovation: an introduction to the concept. *Energy Policy* **35** (5), 2683–2691.

WWAP 2018 *The United Nations World Water Development Report 2018: Nature-Based Solutions for Water.* United Nations World Water Assessment Programme/UN-Water, UNESCO, Paris.

Xing, Y., Jones, P. & Donnison, I. 2017 Characterisation of nature-based solutions for the built environment. *Sustainability* **9** (1), 149.

Zeller, V., Towa, E., Degrez, M. & Achten, W. M. J. 2019 Urban waste flows and their potential for a circular economy model at city-region level. *Waste Management* **83**, 83–94.

Zölch, T., Henze, L., Keilholz, P. & Pauleit, S. 2017 Regulating urban surface runoff through nature-based solutions – an assessment at the micro-scale. *Environmental Research* **157**, 135–144.

First received 30 June 2019; accepted in revised form 1 November 2019. Available online 5 March 2020

IWA Publishing's authorised EU representative for General Product Safety Regulations is Diane D'Arras, 15 rue Duret, 75116 Paris, France, e-mail: safety@iwap.co.uk.

Printed and bound by CPI Group (UK) Ltd, Croydon, CR0 4YY
12/05/2026
02108881-0020